Albert Günther

The GIGANTIC LAND - TORTOISES

Albert Günther

The GIGANTIC LAND - TORTOISES

ISBN/EAN: 9783741118753

Manufactured in Europe, USA, Canada, Australia, Japa

Cover: Foto ©berggeist007 / pixelio.de

Manufactured and distributed by brebook publishing software
(www.brebook.com)

Albert Günther

The GIGANTIC LAND - TORTOISES

THE

GIGANTIC LAND-TORTOISES

(LIVING AND EXTINCT)

IN THE

COLLECTION

OF THE

BRITISH MUSEUM.

BY

ALBERT C. L. G. GÜNTHER,

M.A., M.D., Ph.D., F.R.S.,

KEEPER OF THE DEPARTMENT OF ZOOLOGY.

LONDON

PRINTED BY ORDER OF THE TRUSTEES.

1877.

ALBUM FLAMMAM.

PRINTED BY TAYLOR AND FRANCIS,
RED LION COURT, FLEET STREET.

PREFACE.

AFTER several unsuccessful attempts to unravel the confused history of the Gigantic Land-Tortoises, I had to resume their investigation in the year 1872 on the receipt of a consignment of remains of the extinct Tortoises of the Mascarenes from L. BOUTON, Esq., of Port Louis. Having then ascertained, with the aid of these remains (fragmentary as they were), that the osteology of these animals formed the safest guide to their study, I continued to investigate the subject in that direction, and in 1874 considered my researches sufficiently advanced to begin their publication with a memoir on the Races of the Galapagos Archipelago, which appeared in vol. clxv. of the 'Philosophical Transactions.' Whilst preparing for publication the account of the races inhabiting islands of the Indian Ocean, the following most important materials reached me, by which entirely new facts came to light, and by which I was enabled to correct some of the views expressed in my first paper :—

1. A pair of adult examples, undoubtedly natives of Aldabra, obtained through the mediation of the Hon. Sir ARTHUR H. GORDON, K.C.M.G., then Governor of Mauritius.

2. The entire series of the remains of the Tortoise of Rodriguez, collected for the Royal Society by H. H. SLATER, Esq. (one of the naturalists accompanying the "Transit-of-Venus" Expedition), and transferred by the Council of the Royal Society to the British Museum.

3. A large series of the remains of the extinct Tortoises of Mauritius, collected by the Hon. E. NEWTON, and kindly placed by him in my hands for examination.

4. The examples of the Tortoises of Albemarle and Abingdon Islands, collected by Commander COOKSON, of H.M.S. 'Peterel.'

The results of my examination of these materials were originally embodied in two memoirs—one read before the Royal Society on January 25, 1877, the other before the Zoological Society on February 6th of the same year, both being accepted by those Societies for publication in their 'Transactions.' Thinking it, however, more convenient to the student to obtain a fully illustrated account of these animals in a connected form, I obtained the permission of both Societies to withdraw my papers for that purpose; and as by far the greater part of the specimens form part of the National Collection, the Trustees of the British Museum were pleased to accept my proposal to publish the whole as a volume of the series of publications issued from the Zoological Department.

Finally, I have to express my thanks to the President and Council of the Royal Society for permission to reproduce for this work the thirteen plates illustrating my first paper in the 'Philosophical Transactions.' The majority of these plates are unaltered reprints; but in some alterations have been introduced, viz. in Plates XXX., XXXI., XLVI., & XLVII.

ALBERT GÜNTHER,
Keeper of the Department of Zoology.

British Museum. Dec. 10, 1877.

CONTENTS.

INTRODUCTION.

Nearly all the voyagers of the 16th and 17th centuries who have left accounts of their adventures and discoveries in the Indian and Pacific oceans mention the occurrence, in certain isolated islands or groups of islands, of gigantic Land-Tortoises in countless numbers. The islands on which they met with these animals, although all between the equator and southern tropic, form two most distant zoological stations, widely different in their physical characteristics. One of those stations was the Galapagos Islands; the other comprised Aldabra (on the north-west of Madagascar), and Réunion, Mauritius, and Rodriguez (on the east of that island). But they had this in common, that at the time of their discovery they were uninhabited by man, or even by any larger terrestrial mammal. Not one of these voyagers ever mentions having met with those Tortoises in any other island of the tropics, or in any portion of the Indian continent; and it is not likely that one or the other would not have mentioned the fact if he had seen them in some novel locality. In fact the hardy sailors of that period took the greatest interest in these animals, which were to them a most important article of food. At a time when a voyage now performed in a few weeks took as many months, when every vessel, for defence' sake and from other causes, carried as many people as it was possible to pack into her, when provisions were rudely cured and but few in kind, those tortoises which could be captured in any number with the greatest ease within a few days proved to be a most welcome addition to the stock. The animals could be carried in the hold of a ship or in any other part, without food, for months, and were slaughtered as occasion required, each tortoise yielding, according to size, from 80 to 300 pounds of excellent and wholesome meat. Thus we are informed that ships leaving the Mauritius or the Galapagos used to take upwards of 400 of these animals on board.

When we consider that these helpless creatures lived for ages in perfect security from all enemies, and that nature had endowed them with a most extraordinary degree of longevity[*], so that the individuals of many generations lived simultaneously in their

---

[*] On this point the testimony is unanimous and not to be doubted; in fact all Tortoises are long-lived. The large male Aldabra Tortoise imported for the Trustees of the British Museum from the Seychelles was known to have been kept in the late owner's family in Cerf Island for about 80 years, and is said to have been already of large size at the period when it was brought to the Seychelles. It was still growing at the time of its death in the Zoological Garden of Regent's Park in the present year. Of another example which is living at Colombo

island home, we can well account for the multitudes found by the first visitors to those islands.

LEGUAT (1691) says that "there are such plenty of Land-Turtles in this isle (Rodriguez) that sometimes you see two or three thousand of them in a flock, so that you may go above a hundred paces on their backs." Down to 1740 they continued to be numerous in Mauritius; for GRANT (Hist. Maurit. p. 194) writes in that year, "We possess a great abundance of fowl, as well as both Land- and Sea-Turtle, which are not only a great resource for the supply of our ordinary wants, but serve to barter with the crews of ships who put in here for refreshment in their voyage to India." Yet they appear to have been much more scattered in the larger island than in Rodriguez; and, according to Admiral KEMPENFELT, who visited the latter island in 1761 (see Grant's Maurit. p. 100), small vessels were constantly employed in transporting these animals by thousands to Mauritius for the service of the hospital. Soon, however, their numbers appear to have been rapidly diminished; the old ones were captured by man, the young ones devoured by pigs. Numbers must have succumbed in consequence of the numerous conflagrations by which the rank vegetation of the plains was destroyed to make room for the plantations of the settler. Early in the present century the work of extermination appears to have been accomplished, and there is at present not a single living example either in the Mauritius or in Rodriguez.

Our knowledge of the indigenous fauna of the island of Réunion is still extremely meagre: but although no remains whatever of a large Tortoise from that island are known to exist in any collection, there cannot be any doubt that a gigantic Land-Tortoise once inhabited Réunion, as may be seen from the following historical evidence.

In the continuation of P. G. VERHUFF's 'Voyage into the East Indies' (De Bry, Ind. Orient. Supplementum nonæ Partis. Francof. 1633. fol.) it is stated on p. 23:—
[Anno 1611] "Die 27 [Decembr. mens.] pervenerunt ad Masquerinen*, insulam nempe, 80 miliaribus a Mauritii insula distantem, quæ 16 miliaria circuitu et ambitu suo continet, nec ullis hominibus habitatur, licet ad victum necessarium *Testudines* pisces que et volucres multiplices abunde suppeditet."

In a letter of R. P. BROWN, published in 'Lettres édifiantes et curieuses écrites des Missions étrangères, par quelques Missionaires de la Compagnie de Jésus,' xxx. Recueil, Paris, 1773, 12mo, p. 324, we read that about 60 years ago a party of Frenchmen landed from Madagascar on Réunion, which they found uninhabited, and that for the first two

---

we have equally reliable evidence, as appears from a paragraph in the 'Ceylon Observer,' April 25th, 1870.
' We learn on good authority that the 'Tortoise' exhibited by Mr. Symons, Uplands, the one which is so well known at the Mutwal end of the town, lived in the Uplands compound for between 150 and 200 years. It was sent from Java as a present to one of the Dutch governors here," &c.

A very young example, 7 inches long, sent to me from Aldabra by Dr. W. M'Gregor, was three years old.

* The Portuguese voyager Mascaregnas gave his own name to the island of Bourbon, which formerly had been called by his countrymen *Cerne*.

years they lived almost exclusively on fish and Land- and Marine Tortoises. And again, on p. 338, the writer says:—"L'isle abondoit autrefois en tortues de terre, mais les matelots en ont tant détruit, qu'on n'en trouve plus guère que sur la côte occidentale, encore y sont-elles très-rares. On attribue à ces animaux plusieurs propriétés, entr' autres celle de purifier la masse du sang; et de guérir des maladies. . . . On en tire encore une huile fort douce, qui a presque le même gout que l'huile de Provence" (which statements are repeated in Rees's 'Cyclopædia,' p. 207).

Independent but similar evidence is given by FRANÇOIS CAUCHE in "Relation du Voyage à Madagascar" (1638), published in 'Relations véritables et curieuses,' Paris, 1651, 4to, p. 7. "De là [de l'isle de Diego Rois *] nous tirasmes en l'isle de Mascarhene, qui en est éloignée de 30. lieuës, scituée environ deux degrez delà le Tropique du Capricorne. . . . Elle est inhabitée [uninhabited] comme la précédente, quoy que les eaux y soient bonnes, abondante en gibier, poissons et fruit. On y voit grand nombre d'oiseaux, et *tortues de terre*, et les rivières y sont fort pisqueuses."

Finally I may refer to another testimony in the xvi. Recueil of the 'Lettres édifiantes,' &c. quoted above (Paris, 1724, 12mo), in a "Lettre du Père Jacques," p. 9. "Le meilleur de tous les animaux, qu'on y [in Réunion] trouve, soit pour le goût, soit pour la santé, c'est la *Tortue de terre*. . . . La Tortue est de la même figure que celle, qu'on voit en France; mais elle est bien différente pour sa grandeur. On assure qu'elle vit un temps prodigieux, qu'il lui faut plusieurs siècles pour parvenir à la grosseur naturelle, et qu'elle peut passer plus de six mois sans manger. On en a gardé dans l'isle de petites qui au bout de 20 ans n'avaient grossi que de quelques pouces," &c.

As mentioned above, nothing else is known of this Tortoise: no specimen is known to have been preserved; no remains have been found hitherto. It seems to have been exterminated even before the period of extinction of the Mauritius and Rodriguez species. The Seychelles do not appear to have been inhabited by these animals, certainly not within historical times, all the individuals found there having been imported from Aldabra, and kept in a semi-domesticated condition. The latter group of islands is the only spot in the Indian Ocean where this Chelonian type still lingers in a wild state, in small and gradually diminishing numbers, as we shall subsequently see in the description of the Tortoises of that locality.

The original condition and the fate of the Tortoises of the Galapagos archipelago were precisely the same as in the islands of the Indian Ocean. According to the unanimous testimony of geographers, the first discoverers of those islands, the Spaniards, found them so thickly peopled with Tortoises, that they applied the Spanish word for Tortoise to their discovery. In DAMPIER's time (1680) it was the common practice of vessels to visit those islands for a supply of water and tortoises. In his 'New Voyage round the World' (Lond. 1697, 8vo), p. 101, he says:—"The Land-Turtle are here so numerous that five or six hundred men might subsist on them alone for several months,

* This does not appear to me to have been Rodrigues, as believed by the editor of the voyage.

without any other sort of provision. They are extraordinary large and fat, and so sweet that no pullet eats more pleasantly. One of the largest of these creatures will weigh 150 or 200 weight [pounds], and some of them are 2 foot or 2 foot 6 inches over the callapee or belly [across the sternum]. . . . They have very long small necks and little heads."

The condition of this group of islands, and of the animals inhabiting them, appears to have been unaltered when they were visited by AMASA DELANO and DAVID PORTER—the former a captain in the merchant service, the latter in the navy of the United States.

DELANO ('Narrative of Voyages and Travels,' Boston, 1817, 8vo) made several visits to the Galapagos, the first in 1800 (p. 369). He found plenty of Tortoises in Hood's, Charles, James, and Albemarle Islands. He gives a good description of them, noticing particularly the long, serpent-like neck (p. 376):—" I have seen them with necks between two and three feet long. . . . They would raise their heads as high as they could, their necks being nearly vertical, and advance with their mouths wide open. . . . They are perfectly harmless. . . . I have known them live several months without food; but they allways in that case grow lighter, and their fat diminishes. . . . I carried at one time from James Island 300 very good terrapins to the island of Massa Fuero; and there landed more than half of them, after having them more than 60 days on board my ship. Half of the number landed died as soon as they took food. . . . those that survived the shock which was occasioned by this sudden transition from total abstinence to that of abundance soon became tranquil, and appeared to be as healthy and as contented with the climate as when they were at their native place; and they would probably have lived as long, had they not been killed for food. . . . I have carried them to Canton at two different times."

PORTER informs us of many interesting particulars in his 'Journal of a Cruise made to the Pacific Ocean' (New York, 1822, 8vo, in 2 vols.). He found the Tortoises (in 1813) in greater or less abundance in all the larger islands of the group which he visited, viz. Hood's, Marlborough, James, Charles, and Indefatigable (Porter's) Islands. On Chatham Island, where he made a short stay, a few of their shells and bones were seen, but they appeared to have been long dead (vol. i. p. 231); and on Albemarle Island, the largest of the group, none were observed by him, evidently because he landed here only for a few hours on the south-western point. Abingdon, Binloe, Downe, and Barrington Islands were not visited by him. Some of the Tortoises captured weighed from 300 to 400 pounds (p. 127). "Their steps are slow, regular, and heavy; they carry their body about a foot from the ground. . . . Their neck is from 18 inches to 2 feet in length, and very slender; their head is proportioned to it, and strongly resembles that of a serpent. . . . No animal can possibly afford a more wholesome, luscious, and delicate food than they do. . . . What seems the most extraordinary in this animal is the length of time that it can exist without food; for I have been well assured that they have been piled away among the casks in the hold of a ship, where they have been kept eighteen months, and when killed at the expiration of that time were found to have suffered no diminution in fatness or excellence. They carry with them a constant supply of water in a bag at the root of the neck, which contains about two gallons; and on tasting that found in those we killed on board, it proved perfectly fresh and sweet. . . . In the day-time they appear remarkably quick-sighted and timid, drawing their head into their shell on the slightest motion of any object; but they are entirely destitute of hearing, as the loudest noise, even the firing of a gun, does not seem to alarm them in the slightest degree; and at night, or in the dark, they appear perfectly blind " (p. 150). Near a bay on the north-east part of James Island, Porter took on board as many as would weigh about 14 tons, the individuals averaging about 60 pounds —that is, about 500 individuals (p. 214); and he states that " among the whole only three were male, which may be easily known by their great size, and from the length of their tails, which are much longer than those of the females. As the females were found in low sandy bottoms, and all without exception were full of eggs, of which generally from ten to fourteen were hard, it is presumable that they came down from the mountains for the express purpose of laying. This opinion seems strengthened by the circumstance of there being no male Tortoises among them, the few wo found having been taken a considerable distance up the mountains. . . . The temperature of the air of the Gallipagos Islands varies from 72° to 75°; that of the blood of the Tortoise is always 62°. . . . The eggs are perfectly round, white, and of 2½ inches diameter " (pp. 215, 216).

Very significant are Porter's remarks as regards the differences of the Tortoises from different islands. On Indefatigable Island " they were generally of an enormous size, one of which measured 5½ feet long, 4½ feet wide, and 3 feet thick; and others were found by some of the seamen of a larger size " (p. 164). " The shells of those of James Island are sometimes remarkably thin and easily broken, but more particularly so as they become advanced in age. . . . Those of James Island appear to be a species entirely distinct from those of Hood's and Charles Islands. The form of the shell of the latter is elongated, turning up forward in the manner of a Spanish saddle, of a brown colour, and of considerable thickness. They are very disagreeable to the sight, but far superior to those of James Island in point of fatness; and their livers are considered the greatest delicacy. Those of James Island are round, plump, and black as ebony, some of them handsome to the eye; but their liver is black, hard when cooked," &c. (pp. 214, 215). The Tortoises of Hood's Island were small, similar to those of Charles Island (p. 233).

Before we pass from Porter to his successors, we must mention that he proceeded, after his cruise round the Galapagos, to the Marquesas Islands, making a prolonged stay at Madison Island, where he " distributed from his stock several young Tortoises among the chiefs, and permitted a great many to escape into the bushes and among the grass " (vol. ii. p. 109).

Captain JAMES COLNETT's visit to the Galapagos archipelago deserves to be mentioned only because he adds Abingdon Island to the list of those in which Tortoises occur ('Voyage to the South Atlantic,' Lond. 1798, 4to, p. 152). Also Capt. BASIL HALL landed on this island in 1822, where he found plenty of large Tortoises, of which he laid in a stock which lasted the ship's company for many weeks ('Extracts from a Journal,' Edinb. 1824, 8vo, 2nd edit. vol. ii. p. 140).

Twenty-two years had passed since Porter's cruise when DARWIN visited the Galapagos in the 'Beagle' in the year 1835. A change, by which the existence of these animals was much more threatened than by the casual visits of buccaneers and whalers, had taken place. The Republic of the Equator had taken possession of the archipelago; and a colony of between two and three hundred people banished by the Government had been established on Charles Island, who reduced the number of Tortoises in this island so much that they sent parties to other islands (for instance James) to catch Tortoises and salt their meat ('Journal,' pp. 375, 376). Pigs had multiplied, and were roaming about the woods in a feral state. Darwin adds many interesting observations on the habits of these Tortoises; but as his 'Journal' is in everybody's hands, I quote from his account such parts only as have a special bearing on questions with which we shall have to deal in this treatise. He confirms Porter's observation as regards their deafness, also that "the old males are the largest, the females rarely growing to so great a size. The male can readily be distinguished from the female by the greater length of its tail" (p. 382). An egg which he measured was 7⅜ inches in circumference, a measure nearly identical with that found by Porter. "The old ones seem generally to die from accidents, as from falling down precipices; at least, several of the inhabitants told me they had never found one dead without some evident cause" (p. 384). "The Vice-Governor, Mr. Lawson, declared that the Tortoises differed from the different islands, and that he could with certainty tell from which island any one was brought. . . . M. Bibron, moreover, informs me that he has seen what he considers two distinct species of Tortoise from the Galapagos, but he does not know from which islands. The specimens that I brought from three islands were young ones, and, probably owing to this cause, neither Mr. Gray nor myself could find in them any specific differences ' (p. 394).

After an interval of not quite eleven years, H.M.S. 'Herald' followed the 'Beagle' on a voyage of discovery and survey. The naturalist of that expedition, which reached the Galapagos in the year 1846, found that the progress of civilization had been great ('Narrative of H.M.S. Herald,' by B. Seemann: Lond. 1853, 8vo), or, in other words, that the displacement of the indigenous fauna by man and his companions had proceeded apace. On Charles Island "the cattle had increased wonderfully, and were estimated at 2000 head, beside wild pigs, goats, and dogs. . . . The wild dogs keep the goats and pigs very much down" (vol. i. p. 57); but "no turpin, or terrapin, are living on this island " (p. 59); that is, the Tortoises had been exterminated between the visits of the 'Beagle' and the 'Herald.' On Chatham Island "we saw, for the first time, the terrapin or galapago. . . . We bought them at the rate of six shillings apiece. They were 2 feet 2 inches in length, 1 foot 10 inches broad, standing 1 foot 2 inches off the ground." No specimens were brought home by this expedition.

In the year 1875 the writer of this work directed the attention of Rear-Admiral the Hon. A. A. Cochrane, then Commander-in-Chief on the Pacific Station, to our imperfect knowledge of the present condition of the indigenous fauna of these islands, and especially to the importance of making new and reliable observations on the Tortoises of the various localities. Rear-Admiral Cochrane instructed Commander Cookson, who was then proceeding in H.M.S. 'Peterel' to the archipelago, to obtain, if possible, the desired information and specimens. In this Commander Cookson was most successful, obtaining, besides a series of examples from two of the islands, the following information, which is abstracted from his report to Rear-Admiral Cochrane:—

"These Tortoises are extinct in Charles Island; and only a very few individuals are supposed to survive on Chatham Island. In Hood, James, and Indefatigable Islands the numbers are so reduced that they are no longer hunted, the few left being in the most inaccessible parts of the islands; and I was assured that a search of a fortnight might not result in finding a single individual on either of these islands. Albemarle and Abingdon are the only remaining islands in which they have ever been found. In parts of Albemarle Island they are still very abundant, especially at the south-east end.

"They are still tolerably numerous near Tagus Cove. Landing a party of twenty-four men about half a mile south-east of Tagus Cove, we found in a few hours thirty tortoises; the three largest weighed respectively 241 lb., 185 lb., and 173 lb.; these, I was told, were as large as they are commonly found now.

"Tagus Cove is a favourite resort of whalers for the purpose of getting tortoises. The anchorage is perfectly secure; and the custom is for almost the entire crew to be landed until as many tortoises are secured as can be conveniently taken on board, some whalers going to sea with as many as 100.

"We found a good trail leading from the landing-place (at one of the gullies before mentioned as having pools of fresh water at its mouth) to the ground where tortoises are found, a distance of about three miles; quantities of tortoise-shells and traces of fires showed the numerous camping-grounds.

"Tortoises were never, I believe, very abundant on Abingdon Island: our searching party found four on this island. They were on the high ground; and it was a work of great labour getting them down to the boats. The distance was about four miles; but the ground was exceedingly rugged, and covered with thick brush, through which a trail had to be cut for the entire distance. The largest found on this island weighed 201 lb., and the smallest 135 lb.

"In consequence of the extent of Albemarle Island, and the inaccessibility of many parts of it, I have no doubt these animals are still very numerous on it, and likely to be so for a long period, even at the present rate at which they are destroyed; but I have already shown the havoc made amongst them by the oil-makers. This is the cause of their being nearly extinct on James and Indefatigable Islands, where they used to be so numerous. Admiral Fitzroy found a party on James Island making oil in 1835.

"In Abingdon Island, where they are not numerous, I believe they are doomed to destruction directly the orchilla-pickers are placed on the island; for a party of sixty or eighty men will soon hunt over this small island and discover every individual on it."

This first effort of resuming and supplementing Fitzroy's and Darwin's investigations may, it is to be hoped, encourage others to complete what Commander Cookson had to leave unaccomplished. Hood, James, and South Albemarle are deserving of particular and early attention; and, if no living and adult specimens can be obtained, any remains, especially fragments of carapaces and skulls, from those islands, as well as from Chatham, Indefatigable, and Charles Island (where the Tortoises are said to be quite extinct) will prove of value. It may also be remembered that all the Abingdon Tortoises found were males, and that, beside the skull and cervical vertebræ, we do not know the remaining bones, the entire animal which had been preserved in spirits having been lost during the homeward journey.

H.M.S. 'CHALLENGER' likewise paid a visit to the archipelago immediately after the 'Peterel,' but without obtaining additional information. The Tortoises brought home in this vessel were obtained by, and transshipped from, the 'Peterel.'

The instances of occurrence of the same form of terrestrial animal at widely distant points of the globe are more numerous than we find recorded in treatises on geographical distribution of animals. They form the most interesting element in the investigation of this subject, inasmuch as they have a direct bearing upon, and will greatly assist in the solution of, the problems of the origin of species and the history of the distribution of animals. Of all those instances no one appears to me more remarkable than the reappearance of the " Indian " Gigantic Land-Tortoises in the Galapagos—not in typical singularity, but with nearly all the principal secondary modifications reproduced. How can this be explained with the aid of the doctrine of either a common or manifold origin of animal forms?

On the hypothesis that there is no immediate genetic relationship between the Tortoises of the Galapagos and Mascarenes, we may assume, without overstepping too far the limits of probability, that some terrestrial Tortoises were carried by stream and current, or some other agency, from the American continent to the Galapagos, and that others from Madagascar or Africa found in a similar manner a new home in the Mascarenes. These Tortoises may originally have differed from each other, like the *T. tabulata*, *radiata*, and *sulcata* of our days, possibly not exceeding those species in size; but, being placed under similar external physical conditions evidently most favourable for their further development, they assumed in course of time the same gigantic proportions and other peculiarities, the modifications in their structure which we observe

now being partly genetic, partly adaptive. To test (if test it can be called) this hypothesis, I compared our Tortoises with those of the neighbouring continents, but did not
succeed in finding any evidence in their osseous frame which might be used as
argument in favour of such direct genetic relationship. In none have I found the
singular modification of the articulations of the nuchal vertebræ of the Aldabra
Tortoise.

The naturalists who maintain a common origin for allied species, however distant in
their habitats, will have to assume a former continuity of land (extending over 150° of
longitude, but probably varying in extent, and interrupted at various periods) between
the Mascarenes and Africa, between Africa and South America, and, finally, between
South America and the Galapagos. A continuity of land in this direction is more
probable than one in the opposite hemisphere, which would extend over 210°. Indeed
the terrestrial and freshwater faunæ of Tropical America and Africa offer so many
points of intimate relationship as very strongly to support such a theory. The Tortoises, then, would be assumed to have been spread across the whole of this large area,
without being able long to survive the arrival of man or large carnivorous mammals.
The former, especially before he had provided himself with missile weapons, would
eagerly seek for them, as they were the easiest of his captures, yielding a most
plentiful supply of food; consequently they were exterminated on the continents, only
some remnants being saved by having retired into places which, by submergence, became separated from the mainland before their enemies followed them. The discovery
of bones of Tortoises in Malta, which seem to be almost identical with the corresponding parts in Galapagos Tortoises, as stated in a paper read before the Geological
Society only a short time ago by Dr. Leith Adams, clearly proves the much greater
extension of this Chelonian type at a former geological epoch.

SYNOPSIS

OF THE

EXTINCT AND LIVING GIGANTIC LAND-TORTOISES.

BEFORE proceeding to the detailed description of the various forms, I propose to indicate in a short synopsis the systematic results obtained in the following pages.

I. *Nuchal plate present. Frontal portion of the skull convex. Third cervical vertebra biconvex. Pelvis with narrow symphysial bridge. Gular plate double. Carapace thick.* . . . . ALDABRA TORTOISES.

  1. Shell broad, declivous in front; scutes striated. Skull broad, with the posterior margin of the paroccipital not excised . . . . . 1. *Testudo elephantina*
          (living).

  2. Shell narrow, with horizontal anterior profile; scutes almost entirely smooth. Sternum truncated in front and behind. Skull broad, with the posterior margin of the paroccipital deeply excised . . . . . . 2. *T. daudinii*
          (living?).

  3. Shell regularly elliptical, declivous in front; scutes with shallow striation. Sternum notched behind. Skull narrow, with the posterior margin of the paroccipital not excised; a styloid process on each side of the occipital condyle . .   . . 3. *T. ponderosa*
          (living).

  4. Shell regularly elliptical, declivous in front; scutes quite smooth. Sternum truncated behind. Skull ——? . . . . 4. *T. hololissa*
          (living).

II. *Nuchal plate absent. Frontal portion of the skull flat. Fourth cervical vertebra biconvex. Pelvis with broad symphysial bridge.*

A. *Gular plate single; sternum short* . . . . MASCARENE TORTOISES.

  a. Carapace thin, thickened towards the margins; centre of the last vertebral plate raised into a hump, which is separated from the penultimate vertebral by a transverse depression. . . . . *Tortoises of Mauritius* (extinct).

    Shell depressed, with the upper profile undulated, slightly declivous in front. Lower jaw with a treble dental ridge. Horizontal diameter of pelvis not longer than vertical . . . . 5. *T. triserrata*
          (an = *T. grayi*?).

2. Shell high, with the upper profile convex, declivous in front. Lower
jaw with double dental ridge. Vertical diameter of pelvis longer
than horizontal . . . . 6. *T. inepta.*
3. Shell with the upper profile straight, not declivous in front. (Known
from the shell only.) . . . . 7. *T. indica.*
4. Known from scapula, pelvis, and femur only, the latter bones being
more slender than in 5 and 6 . . . . 8. *T. leptocnemis.*

b. The entire carapace extremely thin; all the bones very slender. *Rodriguez Tortoise* (extinct).

5. Mandible with two dental ridges; the sixth cervical vertebra with a pair
of deep hollows or perforations. Shell not declivous in front in the
male, without vertebral humps. . . . . 9. *T. vosmæri.*

B. *Gular plate double; sternum of moderate extent* . . . . . . GALAPAGOS TORTOISES.

1. Shell broad, depressed, with the upper anterior profile subhorizontal in
the male; plates striated; sternum truncated behind. Outer pterygoid-edge sharp; a deep recess in front of the occipital condyle . . 10. *T. elephantopus*
(James Isl. ? Living).
2. Shell broad, high, with declivous anterior profile in the male; plates
striated; sternum excised behind. Outer pterygoid-edge sharp; a
deep recess in front of the occipital condyle . . . . 11. *T. nigrita*
(Isl. —? Living).
3. Shell rather narrow, depressed, with the upper anterior profile subhorizontal in the male; plates striated; sternum constricted in front and
excised behind. Outer pterygoid-edge flattened; no recess in front
of the occipital condyle. Tympanic opening circular . . . . . 12. *T. vicina*
(South Albemarle ? Living).
4. Shell depressed, with the anterior upper profile straight in the male and
declivous in the female; plates not striated in the adult. Sternum
with short lobes, and truncated behind in the adult. Outer pterygoid-edges flattened, scarcely converging behind. No recess in front
of the occipital condyle . . . . 13. *T. microphyes*
(North Albemarle. Living).
5. Shell elongate, compressed, and not declivous in front, in the male;
plates smooth in the adult; sternum truncated behind. Outer
pterygoid-edges flattened, scarcely convergent; palate shallow. No
recess in front of the occipital condyle . . . . 14. *T. ephippium*
(Indefatigable Isl. Extinct).
6. Shell elongate, compressed, and not declivous in front, in the male;
plates not striated in the adult; sternum truncated behind. Jaws
white. Outer pterygoid-edges flattened, rapidly converging behind.
Palate deeply concave. A deep impression in front of the occipital
condyle . . . . 15. *T. abingdonii*
(Abingdon. Living).

c 2

LIST OF SPECIMENS OF GIGANTIC LAND-TORTOISES, LIVING AND EXTINCT, IN THE
COLLECTION OF THE BRITISH MUSEUM.

1. TESTUDO ELEPHANTINA.

a. Adult ♂, stuffed.		
Carapace 30 in.		74. 2. 6. 3.
b. Skull of *a*.		74. 2. 6. 4.
c. Adult ♂, stuffed.	Aldabra.	Purchased.
Carapace 19 in.		77. 5. 15. 1.
d. Skeleton of *c*.		77. 5. 15. 2.
e. Adult ♀, stuffed.		74. 2. 6. 1.
Carapace 27½ in.		
f. Skull of *e*.		74. 2. 6. 2.
g. Young.		From the Haslar Collection.
Carapace 14 in.		55. 10. 16. 151.
h. Adult ♂.		Purchased of Mr. Bartlett.
Carapace 40 in.		75. 4. 13. 2.
i. Adult ♂.		Old Collection.
Carapace 40 in., without scutes.		77. 4. 28. 1.
k. Skeleton of half-grown ♂.		
Carapace 31 in., sternum missing.		74. 2. 6. 8.
l. Half-grown ♂.		From the R. Coll. of Surgeons.
Carapace 29½ in.		76. 1. 14. 2.
m. Very young, in spirits.	Aldabra.	Pres. by Col. Playfair.
		77. 5. 12. 1.
n. Very young, stuffed.		Pres. by General Hardwicke.
o. Skull of adult ♂.		16. *b*.
p. Skull of half-grown spec.		16. *a*.
q. Two humeri of adult.		Purchased. 76. 11. 21. 2-3.
r. Scapula of adult.		Purchased. 76. 11. 21. 1.
s. Femur of adult.		Purchased. 76. 11. 21. 4.
t. Young, malformed.		Purch. of the Zool. Soc.
Carapace 12½ in.		

2. Testudo daudinii.

a. Adult ♂, stuffed.		
Carapace 34½ in.		7 4. 2. 6. 5.
b. Skull of a.		7 4. 2. 6. 6.
c. Bones of forearms and lower legs of a.		74. 2. 6. 6.

3. Testudo ponderosa.

a. Adult ♀, skeleton.		Purchased. 64. 12. 20. 27.
Carapace 27½ in.		
b. Pelvis of adult.		Purchased. 76. 1. 4. 1.

4. Testudo hololissa.

a. Adult ♀, living.	Aldabra.	Purchased.
Carapace 32 in.		

5. Testudo triserrata.

a. Skull of adult.	Mare aux Songes.	Pres. by E. Newton, Esq.
b. Mandible.	,,	Pres. by Col. A. E. H. Anson. 39929.
c. Fragment of cranium.	,,	Ditto. 39928.
d. Fragment of carapace.	,,	Ditto. 39964.
e. Sternum.	,,	Pres. by E. Newton, Esq.
f. Two left scapularies.	,,	Pres. by Col. A. E. H. Anson. 39939 & 39941.
g. One right scapulary.	,,	Ditto. 39940.
h. Four humeri.	,,	Ditto. 39942, 39945–47.
i. Four pelves.	,,	Ditto. 39932–35.

6. Testudo inepta.

a. Skull of adult.	Mare aux Songes.	Pres. by Col. A. E. H. Anson. 39927
b. Skull of adult.	,,	Pres. by E. Newton, Esq.
c. Carapace without sternum.	,,	Pres. by L. Bouton, Esq.
d. Fragment of carapace with part of first rib.	,,	Pres. by Col. A. E. H. Anson. 39963.
e. Right scapulary.	,,	Ditto. 39937.

f. Left scapulary.	Mare aux Songes.	Pres. by Col. A. E. H. Anson. 39938.
g. Two left humeri.	„	Ditto. 39943–44.
h. Pelvis.	„	Ditto. 39931.
i. Pelvis.	„	Purchased of Mr. Bewsher. 75. 10. 20. 1.

7. TESTUDO LEPTOCNEMIS.

a. Two right scapularies.	District of Flacq.	Pres. by L. Bouton, Esq. 76. 11. 4. 12–13.
b. Four fragments of pelvis.	„	Ditto. 76. 11. 4. 14–17.
c. Two left femurs.	„	Ditto. 76. 11. 4. 10–11.

* SPECIES INCERTA.

a. Fifth cervical vertebra.	Mare aux Songes.	Pres. by Col. A. E. H. Anson. 39930.
b. Left scapula (" *T. boutonii* ")	„	Ditto. 39936.
c. Left ulna.	„	Ditto. 39948.
d. Right ulna.	„	Ditto. 39949.
e. Left radius.	„	Ditto. 39951.
f. Three right radii.	„	Ditto. 39950, 39952–3.
g. Three left femurs.	„	Ditto. 39954–6.
h. Three right femurs.	„	Ditto. 39957–9.
i. Left tibia.	„	Ditto. 39960.
k. Right tibia.	„	Ditto. 39961.
l. Right fibula.	„	Ditto. 39962.

8. TESTUDO VOSMÆRI.

All the specimens are from Rodriguez.

a. Skulls and fragments of skulls of various sizes.		" Transit-of-Venus " Collection. 76. 11. 1. 1–20.
b. More or less incomplete sets of cervical vertebræ of various sizes.		Ditto. 76. 11. 1. 21–49.
c. Fifth cervical vertebra (fig.).		Ditto. 76. 11. 1. 117.
d. A set of three cervical vertebræ (fig.).		Ditto. 76. 11. 1. 118.

e. Fifth and seventh cervical vertebræ. Pres. by L. Bouton, Esq. 76. 11. 4. 6-7.

f. Sixth and seventh cervical vertebræ. Ditto. 76. 11. 4. 1-2.

g. Eleven dorsal vertebræ. " Transit-of-Venus " Collection. 76. 11. 1. 51.

h. Seventeen caudal vertebræ. Ditto. 76. 11. 1. 50.

i. Five carapaces of half-grown and young individuals. Ditto. 76. 11. 1. 52-56.

k. Six pairs of scapulæ and several odd ones of various sizes. Ditto. 76. 11. 1. 57-64.

l. One scapula (fig.). Ditto. 76. 11. 1. 115.

m. Another scapula. Pres. by L. Bouton, Esq. 76. 11. 4. 9.

n. Three pairs of humeri and several odd ones of various sizes. " Transit-of-Venus " Collection. 76. 11. 1. 65-68.

o. A humerus (fig.). Ditto. 76. 11. 1. 114.

p. Two humeri. Pres. by L. Bouton, Esq. 76. 11. 4. 8.

q. Several right and left ulnæ. " Transit-of-Venus " Collection. 76. 11. 1. 69-71.

r. Two other ulnæ (fig.). Ditto. 76. 11. 1. 120.

s. Several right and left radii. Ditto. 76. 11. 1. 72-76.

t. Pelves of various sizes. Ditto. 76. 11. 1. 77-98.

u. Two other pelves (fig.). Ditto. 76. 11. 1. 113.

v. Two other pelves. Pres. by L. Bouton, Esq. 76. 11. 4. 4-5.

w. Several right and left femurs. " Transit-of-Venus " Collection. 76. 11. 1. 99-100.

x. Two other femurs (fig.). Ditto. 76. 11. 1. 116.

y. Several right and left tibiæ. Ditto. 76. 11. 1. 101-104.

z. Right tibia. Pres. by L. Bouton, Esq. 76. 11. 4. 3.

α. Several right and left fibulæ. " Transit-of-Venus " Collection. 76. 11. 1. 105-108.

β. Another fibula (fig.). Ditto. 76. 11. 1. 119.

γ. Two metatarsal bones. Ditto. 76. 11. 1. 109.

δ. Nine phalanges. Ditto. 76. 11. 1. 110.

9. Testudo elephantopus.

a. Skeleton of young.   Galapagos.   Pres. by Capt. E. M. Leeds.
76. 10. 23. 1.

10. Testudo nigrita.

a. Adult ♂.   Purchased.
    Carapace 41 in. (without sternum).

b. Skeleton of young.   Galapagos.   Presented by Dr. Günther.
    Carapace 16 in. long.   76. 10. 23. 2.

c. Skeleton of young.
    Carapace 15½ in. long.   47. 3. 5. 27.

d. Young, stuffed.   From the Haslar Collection.
    Carapace 10 in. long.

e. Young, stuffed.   Galapagos.   Purchased. 70. 12. 18. 1.
    (See Proc. Zool. Soc. 1870, pl. xli.)

f–h. Very young, stuffed.
    Carapaces 6, 5½, and 4½ in. long.

i. Very young, shell.   Pres. by Dr. Gray.
    Carapace 3¼ in. long.

k. Skull of adult.   From the Haslar Collection.
    (*T. planiceps*, Gray.)   48. 4. 6. 3.

11. Testudo vicina.

a. Skeleton of adult.   Obtained by exchange.
74. 7. 15. 1.

b. Skull of adult.   Iguana Cove, S. Albemarle.   Commander Cookson's Collection.
74. 6. 21. 43.

c. Young, stuffed.   74. 6. 1. 4.
    Carapace 12½ in.

d. Skull of *c.*   74. 6. 1. 5.

12. Testudo microphyes.

a. Adult ♂ (still living).   Tagus Cove, North   Commander Cookson's Collection.
    Carapace 33¼ in.   Albemarle.

b. Adult ♂.   ,,   Ditto. 76. 6. 21. 41.
    Carapace 33¼ in.

c. Skull of *b.*   ,,   Ditto. 76. 6. 21. 41.

d. Half-grown ♂. Carapace 25 in.	Tagus Cove, North Albemarle.	Commander Cookson's Collection. 76. 6. 21. 42.
e. Skull of *d.*	,,	Ditto. 76. 6. 21. 42.
f. Adult ♀ (still living). Carapace 27 in.	,,	Ditto.
g. Adult ♀, stuffed. Type of the species.		Purchased. 75. 12. 29. 1.
h. Skull of *g.*		75. 12. 29. 1.
i. Young (still living). Carapace.	Tagus Cove, North Albemarle.	Commander Cookson's Collection.

13. TESTUDO EPHIPPIUM.

a. Young, stuffed. Carapace 7 in.		74. 6. 1. 6.

14. TESTUDO ABINGDONII.

a. Adult ♂. Carapace 38 in.	Abingdon Island.	Commander Cookson's Collection. 76. 6. 21. 40.
b. Skull of *a.*		
c. Adult ♂, stuffed. Carapace 34 in.	,,	Ditto. 76. 6. 21. 39.
d. Skull of *c.*		Ditto. 76. 6. 21. 39.
e. Adult ♂, stuffed. Carapace 38¼ in.	,,	Ditto. 76. 6. 21. 38.
f. Skull of *e.*		Ditto. 76. 6. 21. 38.
g. Cervical vertebræ of *e.*		Ditto. 76. 6. 21. 44.

D

THE RACES OF THE ALDABRA GROUP.

ALL the living gigantic Tortoises which have been brought to Europe from the Indian Ocean within the last forty years are distinguished by the presence of a nuchal shield, convex cranium, and narrow obturatoric bridge. Whenever it has been possible to trace their origin this was found to be Aldabra, a group of small islands north-west of Madagascar, situated in lat. 9° 25′ S., long. 46° 20′ E. Although the majority of the specimens were shipped at Mauritius or the Seychelles, they had been brought there, in the first instance, from Aldabra, to be kept as curiosities or for the sake of their flesh or that of their progeny, which, in that suitable climate, they annually yielded to their owners with great regularity.

It is not at all impossible, nay, probable, that some of the other insignificant islands which are scattered in the same part of the Indian Ocean likewise harboured this Chelonian type; but of this I cannot find anywhere in the old records positive evidence. If it was so, the Tortoises were speedily exterminated, as the islands, from their small size, could not offer to them any place of effectual concealment. These islands seem to have been formerly much more frequently visited by ships on their way to or from India than at present, when they are rather avoided.

Among the specimens united by the common characters stated above, not unimportant differences may be observed; they are difficult to reconcile with our ideas about the amount of variability within the limits of the same species or race; but we can account for them by assuming that the animals came from different though neighbouring islands. More especially the islands of which the Aldabra group consists are separated by narrow deep channels of the sea, perfectly impassable to animals so helpless in the water as Land-Tortoises. In other respects the islands are described as covered with verdure, low tangled bushes interspersed with patches of white sand. Two of the islands are rather low, hummocky near the centre. The third island is the largest, extending about eighteen sea-miles in length and two or four miles in breadth; it is much higher than the others, and partly covered with very high trees, that may be seen eight or nine leagues from the deck of a moderate-sized ship. Besides these three larger islands, there are several very small outlying islets. The first record of the existence of Land-Tortoises in Aldabra I find on a map of this group given by

Dalrymple from the observations made in the boat 'Charles' and tartan 'Elizabeth' in 1744, where, the author adds, " they found a great many Land-turtle much larger than those at Rodrigue." Our knowledge of the configuration of these islands is still so incomplete that it will be useful to reproduce Dalrymple's chart; it is clearly faulty in its north-eastern portion (where the channel between the second and third main islands is omitted), but gives some idea of the smaller islets in the lagoon.

ALDABRA (from Dalrymple).

On applying to the Hydrographer, Capt. Evans, C.B., for more recent information, he, with his usual kindness, directed my attention to an unpublished chart deposited by the late Admiral W. F. W. Owen in the archives of the Admiralty, and reproduced here (p. 20) with the permission of the hydrographer. It appears that the group was partially surveyed by Lieut. R. Owen, of H.M.S. 'Leven,' in 1824, which, with H.M.S. 'Barracouta,' was engaged at that time, under the command of Capt. W. F. W. Owen, in the exploration of this part of the Indian Ocean*. This is at present the most reliable

* See 'Narrative of Voyages to Explore the Shores of Africa, Arabia, and Madagascar, performed in H.M. ships " Leven " and " Barracouta," under the direction of Capt. W. F. W. Owen, R.N.' 2 vols. 8vo. Lond. 1833. No complete and continuous detailed account of the work of these ships is given in that 'Narrative;' and more especially I am unable to find Lieut. Owen's visit to Aldabra mentioned.

chart of the group; and the parts of the coast-line actually seen and surveyed by Lieut. Owen are distinctly shown.   As he gives the depth of the lagoon, he must have entered

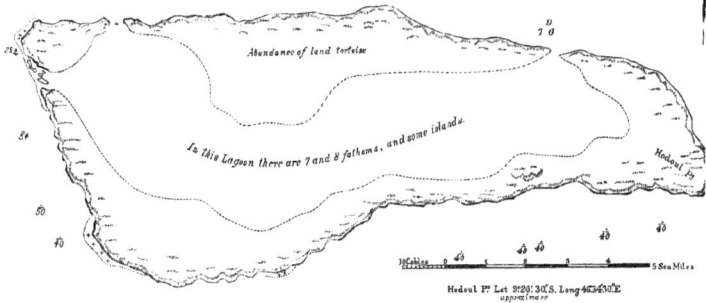

ALDABRA.
(Surveyed by Lieut. Rich. Owen, R.N., H.M.S. 'Leven,' 1824.)

it in boats, although he does not take further notice of the smaller islands, the surveying of which could not possibly be of any importance to the navigator.

Aldabra has never been permanently inhabited—a circumstance by which the Tortoises were saved from utter extinction. Although the number was constantly being thinned by the raids of crews of passing whalers or fishing-vessels, a small remnant found a safe refuge in the tangled and almost impenetrable thickets; if these be allowed to be invaded or cut down, the Aldabra Tortoise will disappear, like those of the Mascarenes *.

* A report having reached England that a permanent settlement was about to be made on Aldabra, steps were taken to secure the protection of the Tortoises; and this will not be an inappropriate place to put on record a memorial sent to the Government of Mauritius, as well as the replies thereto:—

"To His Excellency the Hon. Sir Arthur Hamilton Gordon, K.C.M.G., Governor and Commander-in-Chief of Mauritius and its dependencies.

"We, the undersigned, respectfully beg to call the attention of the Colonial Government of Mauritius to the imminent extermination of the Gigantic Land-Tortoises of the Mascarenes, commonly called 'Indian Tortoises.'

"2. These animals were formerly abundant in the Mauritius, Réunion, Rodriguez, and perhaps other islands of the western part of the Indian Ocean. Being highly esteemed as food, easy of capture and transport, they formed for many years a staple supply to ships touching at those islands for refreshment.

1. TESTUDO ELEPHANTINA.

After having carefully studied the detailed description given by Duméril and Bibron under the above heading, it seems to me very probable that several specifically distinct forms were included in it. The scutes are described by the French herpetologists as sometimes striated, sometimes entirely smooth, whilst I exclude from this species all individuals with perfectly smooth scutes, the absence of sculpture not being dependent either on sex or entirely on age. On the other hand I refer to this species specimens of black as well as brown coloration, whilst the latter is stated to be one of the characteristics of *T. elephantina* in the 'Erpétologie générale.' Thus the limits assigned to this species by its authors do not appear to coincide with those determined by myself.

" 3. No means being taken for their protection, they have become extinct in nearly all these islands; and Aldabra is now the only locality where the last remains of this animal form are known to exist in a state of nature.

" 4. We have been informed that the Government of Mauritius have granted a concession of Aldabra to parties who intend to cut the timber on this island. If this project be carried out, or if otherwise the island is occupied, it is to be feared, nay, certain, that all the tortoises remaining in this limited area will be destroyed by the workmen employed.

" 5. We would therefore earnestly submit it to the consideration of your Excellency whether it would not be practicable that the Government of Mauritius should cause as many of these animals as possible to be collected before the wood-cutting parties or others land, with the view of their being transferred to the Mauritius or the Seychelles Islands, where they might be deposited in some enclosed ground or park belonging to the Government, and protected as property of the Colony.

" 6. In support of the statement above made and the plan now submitted to the Mauritius Government, the following passages may be quoted from Grant's 'History of Mauritius,' 1801, 4to :—

" ' We (in Mauritius) possess a great abundance of both land- and sea-turtle, which are not only a great resource for the supply of our ordinary wants, but serve to barter with the crews of ships' (p. 194).

" ' The best production of Rodriguez is the land-turtle, which is in great abundance. Small vessels are constantly employed in transporting them by thousands to the Isle of Mauritius for the service of the hospital ' (p. 100).

" ' The principal point of view (in Rodriguez) is, first, the French Governor's house, or rather that of the Superintendent appointed by the Governor of the Isle of France to direct the cultivation of the gardens there and to overlook the park of land-turtles ; secondly, the park of land-turtles, which is on the sea-shore facing the house ' (p. 101).

" 7. The rescue and protection of these animals is, however, recommended to the Colonial Government less on account of their utility (which now-a-days might be questioned in consideration of their diminished number, reduced size, and slow growth, and of the greatly improved system of provisioning ships, which renders the crews independent of such casual assistance) than on account of the great scientific interest attached to them. With the exception of a similar tortoise in the Galapagos Islands (now also fast disappearing), that of the Mascarenes is the only surviving link reminding us of those still more gigantic forms which once inhabited the continent of India in a past geological age. It is one of the few remnants of a curious group of animals once existing on a large submerged continent, of which the Mascarenes formed the highest points. It flourished with the Dodo and Solitaire; and whilst it is a matter of lasting regret that not even a few individuals of these curious birds should have had a chance of surviving the lawless and disturbed condition of

The amount of variation in the form of the carapace is considerable, much greater than in any of the other Gigantic Land-Tortoises; and the differences between these variations are quite equal, and in some measure analogous, to those observed between some of the Galapagos Tortoises for which I have adopted distinct specific denominations. But, whilst in the latter such external modifications are combined with marked features of the skull and other parts of the skeleton, the skulls of the varieties of *T. elephantina* surprisingly agree in the details of form with one another.

It is this species to which J. E. Gray and others applied the name of *Testudo indica* par excellence; finding that the varieties of external form closely resemble, and almost repeat, those from the Galapagos, they included in it, more or less, all the other Gigantic Land-Tortoises. That name must be restricted to the species described by Perrault (Mém. pour servir à l'hist. nat. des anim. 1676, p. 103, c. tab.).

Characteristic of this species is the concentric striation of the scutes, which, most distinct in young specimens, never disappears entirely, not even in the largest and oldest examples; then the declivity of the first vertebral scute; the moderate reversion of the margins of the carapace; the undivided caudal scute*; the broad skull, with the posterior margin of the paroccipital not excised.

The specimens available for examination may be grouped as follows:—

---

past centuries, it is confidently hoped that the present Government and people, who support the 'Natural-History Society of Mauritius,' will find a means of saving the last examples of a contemporary of the Dodo and Solitaire.

(Signed)   "Joseph D. Hooker.
              "H. B. E. Frere.
              "Rich. Owen.
              "Charles Darwin.
              "Alfred Newton.
              "John Kirk.
              "Albert Günther."

"London, April 1874."

This memorial received the full attention of Sir A. Gordon, who, in his reply, promised that the lessee should be bound to protect the animals, and to remit annually to Mauritius a pair of living ones, which, with others acquired by purchase, would be preserved in a paddock of the Botanic Gardens at Pamplemousses. On Sir A. Gordon leaving Mauritius soon afterwards, M. Bouton, the Secretary of the Royal Society of Mauritius, submitted the matter to his successor, Sir A. P. Phayre, who replied as follows:—

"My dear M. Bouton,—I return you the papers received from you regarding the gigantic Land-Tortoises. I have made arrangements with the Chief Civil Commissioner in order to have those on Aldabra preserved.

"Believe me,
"Yours sincerely,
(Signed)   "A. P. Phayre."

* But for this character I should regard *Testudo gigantea* of Schweigger and Duméril and Bibron as identical with *T. elephantina*. Possibly the division of this scute is only an individual peculiarity of the specimen on which this species was founded.

A. Each vertebral scute is raised into a high protuberance, more or less confined to the areolar region.

   α. Carapace broad, black or of a dark brown colour.

   β. Carapace narrower, horn-coloured.

B. Vertebral scutes not raised in the middle, flat.

   γ. Carapace broad, black.

Of these three varieties α may be compared to *T. nigrita*, β to *T. vicina*, and γ to *T. elephantopus* from the Galapagos Islands.

The materials which I refer to *T. elephantina* are the following:—

Var. α.

a. An adult stuffed male, carapace 30 inches long, in the British Museum. Black.

b. Skull of the same individual.

c. An adult male, carapace 40 inches long, from Aldabra, in the British Museum. Brown. This specimen is one of the pair imported in 1875 from the Seychelles (see 'Nature,' 1875, p. 260).

d. Skeleton of the same individual.

e. An adult stuffed female, carapace 27¼ inches long, in the British Museum. Black.

f. Skull of the same individual.

g. Carapace of a young specimen, 14 inches long, in the British Museum.

Var. β.

h. Carapace of adult male, 40 inches long, in the British Museum.

Var. γ.

i¹. Carapace of an adult male, without scutes, 40 inches long, transferred from the Old Collection in Montague House (with cast of head and feet).

i². An adult stuffed male, carapace 43 inches long, in the Free Public Museum, Liverpool. The Curator, Mr. T. J. Moore, has kindly informed me that this specimen was brought alive from the Cape of Good Hope by a collector sent about the year 1840 by the then Earl of Derby to South Africa to collect living antelopes &c. It lived for a few months only at Knowsley.

k. Skull, bones of the forearm and lower leg, with the majority of the carpals and tarsals of the same individual.

l. Skeleton of a half-grown male (sternum missing), carapace 31 inches long, in the British Museum.

m. Carapace (with sternum) of a half-grown male, 29¼ inches long, in the British Museum.

Finally—

n. A skull, humerus, femur, and scapula of one or more large examples in the British

Museum cannot be referred to either of these three varieties, nothing being known of the carapaces or any other part of the animals to which they belonged.

o. Skull of a half-grown example in the British Museum, figured in Gray's 'Catal. of Shield-Reptiles' (4to), pl. 35. fig. 1.

Variety α.

In the *adult male* specimen the carapace (Plate III.) is very high, very convex in every direction, and nearly equally broad in front and behind. The anterior vertebral scute slopes rapidly from a protuberance near its hind margin down to the nuchal plate. The anterior and posterior free margins of the carapace are somewhat turned outwards, but not reverted, with shallow notches between the marginal scutes. The caudal plate is neither bent inwards nor outwards, its surface being a plane. The middle of each of the four posterior vertebrals is raised into a high hump, which gradually slopes down towards the margins. Also the costals participate in some measure in this peculiarity; but their most prominent part has an eccentric position, and is limited to the upper half of each scute. In all the scutes the smooth areolar portion can be clearly distinguished from the very broad striated portion, although the striæ are rather broad and somewhat obsolete in the vertebrals, but well-developed in the marginals. Nuchal plate of moderate size, rather longer than broad. The *sternum* (Plate IV. fig. A) is deeply concave, broad, with the anterior portion somewhat contracted, truncated in front and behind. The shell is very thick and heavy *, the substance of the caudal plates and of the lateral portion of the abdominals being particularly thickened and callous. In the large specimen, c, a section of this portion shows a thickness of three inches— the osseous substance, however, not being solid, but cavernous.

The upper surface of the *head* (Plates I. & II.) is covered with large irregular shields, of which two, covering the frontal region, are by far the largest, about thrice as long as broad. The neck is shorter than in any of the Galapagos Tortoises. The scutes on the anterior side of the fore leg are separated from the smaller ones covering the hinder side by a longitudinal series of large scutes, which run along the outer edge of the forearm. The tail is about 10 inches long, and provided with a large, flat, terminal claw.

* FALCONER, in his notes on *Colossochelys atlas* (Palæontolog. Mem. vol. i. p. 378), states that " the thickness of bone in the convexity is almost in an inverse ratio to the size. The physiological reason of this is, that the smaller the animal, the more liable it is to injury, and it requires a greater arch to sustain it." This view is not confirmed by an examination of the living Tortoises; the Aldabra species is as large as those from the Galapagos, and even larger than one of those latter, yet it has a much thicker shell. We shall see that the extinct Mascarene species agree with the Galapagos Tortoises in this respect. Perhaps the cause of this is to be sought in the small quantity of earthy matter contained in the food on which these animals chiefly subsist. A living Galapagos Tortoise in my possession preferred the petals of a *Wrteria* to every other plant. Of course, by the thinness of the shell its weight is much reduced ; and these Tortoises are therefore able to walk faster and to carry the shell higher above the ground than the thick-shelled species. The thinness of the shell and the slender osseous framework of the limbs are, in fact, characters correlated to each other.

The *female* (spec. *e*) does not differ much from the male as regards the form of the carapace; but, being evidently a much younger specimen, the striæ are deeper, more numerous, and more distinct. The nuchal plate is, exceptionally, absent. The sternum is nearly flat, with a distinct notch in front and behind. Tail short, with a terminal claw. Colour as black as in the male.

The *very young* specimen, which is only 14 inches long, has the scutes very neatly sculptured, deep concentric striæ occupying nearly the whole of the scutes, with the exception of the small areolar spaces. The carapace itself is high, of a regularly oval shape. Sternum scarcely notched in front and behind. Colour less intensely black than in the preceding specimens.

The measurements of these specimens are as follows:—

	Length of carapace. In str. line. in.	Length of carapace. Over curve. in.	Width of carapace. In str. line. in.	Width of carapace. Over curve. in.	Depth of carapace. in.	Sternum. Length. in.	Sternum. Width. in.	Caudal plate. Length. in.	Caudal plate. Width. in.
Ad. ♂	49	64½	36½	65	26	38	25	4	8½
Ad. ♂	39	52	26	54	24	32	23	3½	9
Ad. ♀	27½	36	22	37	13½	23½	17½	3½	6½
Yg.	14	18½	8	18	8½	12½	8½	1½	3¼

Variety β.

Of this variety I have only one incomplete example before me (if, indeed, the very young specimen just described should not be more properly referred to it)—a large carapace 40 inches long, the sternum of which is lost. It differs from the preceding variety in being horn-coloured, and in having the carapace conspicuously narrower and somewhat less elevated. The protuberances of the costal plates are less pronounced; and the upper profile of the last vertebral and caudal sweeps downwards in an uninterrupted gentle curve. The broad areolar portions and the striations of the scutes are exactly as in the preceding variety; nuchal plate of moderate size, as long as broad.

	Length of carapace. In str. line. in.	Length of carapace. Over curve. in.	Width of carapace. In str. line. in.	Width of carapace. Over curve. in.	Depth of carapace. in.	Caudal plate. Length. in.	Caudal plate. Width. in.
Ad. ♂	40	50½	23½	40	16½	4	9

Variety γ.

The *adult male* has the carapace broad and depressed, nearly equally wide in front and behind; its upper surface is remarkably flat, and the anterior slope considerably steeper than the posterior; in fact the median profile of the first vertebral scute between its hind margin and the nuchal plate is remarkably concave. The anterior free margin of the carapace is somewhat turned outwards, but not reverted, and without any notches, whilst the posterior shows a reversion with shallow notches between the marginal scutes.

E

Also the caudal plate is slightly bent outwards, with a rather concave surface. The high humps of the vertebral and costal scutes, which are so conspicuous in variety α, are here entirely absent; but in all the scutes the smooth areolar portion can be clearly distinguished from the very broad striated portion, the striæ being rather wide. The sternum is deeply concave, broad, with the anterior portion not conspicuously contracted, and with a shallow notch in front and behind, the substance of the caudal plates and of the lateral portion of the abdominals being thickened and callous. In the scutellation of the head and limbs this variety does not differ from variety α; also the tail is nearly of the same length and provided with a flat terminal claw.

The carapaces of the *younger* specimens agree in the essential particulars with the larger; but they are decidedly higher, the surface of the costal plates being more vertical and the slope of the anterior vertebral scute less steep.

The colour of the three specimens is black, with a horny tinge about the areolar portions.

The measurements of these specimens are as follows:—

	Length of carapace.		Width of carapace.		Depth of carapace.	Sternum.		Caudal plate.	
	In str. line.	Over curve.	In str. line.	Over curve.		Length.	Width.	Length.	Width.
	in.	in.	in.	in.	in.	in.	in.	in.	in.
Ad. ♂	43	50½	30	50	20½	31½	26	3½	9
Ad. ♂	40	52	30	52	18½	32	24	[?]	10
H.-gr. ♂ ...	31	41	19	40	17	[?]	18	3	7
H.-gr. ♂ ...	29½	38	18	40	15	23	17	2¾	7

Osteology.—In the preceding remarks, as well as in the following notes on the osteological characters of the various species, it is not my intention to give such a complete description as would include every detail common to all the species of *Testudo*, but I shall limit myself to those points only by which the various species of gigantic Tortoises differ from one another in a marked manner.

The *skull*, figured on Plates VIII. & IX. fig. A, measures in length from the occipital condyle to the extremity of the snout 5½ inches, and from the extremity of the occipital crest to that of the snout 6¼ inches; its greatest width is 4½ inches. Large as this skull is, it is surpassed in size by that of specimen *c*, the corresponding measurements of which are 6, 6½, and 4½ inches. In the details of their structure and configuration these skulls are perfectly identical. They are distinguished from those of the other species by the very shallow excision of the hind margin of the paroccipital, and by their great relative breadth. 1*. The frontal region is raised and convex, broad, its greatest width (in front of the postfrontal) being one half of the distance of

* In this and the following description of the skulls the same points are noticed under the same numbers—a plan by which the comparison of the several parts (sometimes described many pages apart) is much facilitated. The significance of certain modifications of structure noticed here will become more apparent when we shall treat of the skulls of the Mascarene and Galapagos Tortoises.

the tympanic condyles (outside measure). 2. The upper surface of the parietals is a flat triangle, the outer edges of which converge towards, and finally are merged into the tetrahedral occipital crest, which is of moderate length. 3. The tympanic case with the mastoid is not produced backwards; so that the posterior margin of the paroccipital is but little excised. 4. The groove on the lower surface of the occipital, in front of the condyle, is very shallow; no styloid process on each side of the condyle. 5. The prominent tuberosity on the front margin of the temporal fossa, which in some of the Galapagos Tortoises is so much developed, is but feebly indicated. 6 *. The tympanic cavity is of moderate extent, the outer tympanic rim having an irregular outline, like that of an inverted human concha, the convexity being anterior. In its posterior margin there is a very deep triangular excision for the passage of the Eustachian tube. 7. This excision is so deep that it nearly extends to that part of the bottom of the cavity which is pierced by the foramen for the columella; this ossicle, therefore, does not rest on a trenchant ridge as in the Galapagos Tortoises. 8. The front margin of the intermaxillaries projects so much beyond that of the frontals, that the nasal opening slopes obliquely downwards and is much higher than broad. 9. The inner nostrils are situated far back, and, on the palatal view of the skull, may be seen free and uncovered by the alveolar lamellæ of the maxillaries. 10. The intermaxillaries are long, about two thirds the length of the maxillaries; and their foremost portion is neither very concave nor much bent downwards to form the beak. The suture between the intermaxillary and vomer is at a considerable distance behind the inner angle of the alveolar edges of the maxillaries. 11. The whole of the palatal region is deeply concave, with a very low median crest along its bottom. The triangular space of which the foramina palatina and the anterior extremity of the vomer form the points is isosceles. 12. Anterior surface of the tympanic pedicle deeply excavated. 13. Lower jaw with a double alveolar ridge, the symphysial portion being simply vertical, without a backward dilatation of the lower margin of the bone.

The description of the other parts of the skeleton is taken principally from the adult male specimen *c* and from *l*, a male not fully grown, with a carapace 31 inches long.

The *cervical portion of the vertebral column* is comparatively much shorter than that of the Galapagos or Rodriguez Tortoises; the several vertebræ are shorter and thicker; the crests are lower, without additional lateral crests; the majority of the posterior articulary processes have a strongly divergent outward direction, the notches between them being very wide and deep; and the anterior are nearly perpendicular to the

---

* It is very singular that although the osseous parts of the auditory organ are so well developed, nevertheless, according to the unanimous testimony of observers, these Tortoises are absolutely deaf. I find this confirmed so far in my young living example, that it never takes notice of the noisy approach of a person whom it cannot see, nor is it disturbed by the fall of a stone behind its back. Perhaps the faculty of hearing, although never acute, is not entirely lost until the individuals have attained to a great age.

longitudinal axis of the vertebræ. ˙ All these characters are indicative of a comparatively shorter neck with lessened power of turning, as we actually observe it in the
living animals, although it must be remembered that the length and mobility of the
neck of these Aldabra Tortoises is still much greater than in any of the smaller forms
of the genus *Testudo*. Most remarkable is the peculiarity in the articulation of the
third and fourth vertebræ, which in all the Flat-headed Tortoises (from the Galapagos
as well as Mauritius and Rodriguez) is reversed.

In the *atlas* of the younger individual the lateral portion of the neural arch (column)
is very broad, considerably broader than the zygapophysis, which is rather shorter than
that part of the bone which forms the roof of the arch. These parts are separated
from one another by less-deep constrictions in the adult specimens, the lateral portion
being nearly as broad as the zygapophysis (Plate X. fig. A). The centrum (odontoid
process) is as broad as long in the adult, and broader than long in the young, with
oblique articular surfaces, two in front and one behind.

In the *second* vertebra (Plate X. fig. B) the neural arch is somewhat compressed,
provided with a low crest. The *third* has a condyle in front and behind, the posterior
being much elongate and reaching beyond the extremity of the zygapophyses (Plate X.
fig. C). The *fourth* has a glenoid cavity in front and a condyle behind ; there is a
marked hollow in the neural arch behind the anterior zygapophysis in the adult specimen, which is absent in the young, clearly showing that this peculiarity is developed
with age. The *fifth* has a single low median neural crest (Plate XI. fig. A), and a deep
hollow behind each anterior zygapophysis. The *sixth* has a glenoid cavity in front, and
a broad, depressed, bipartite condyle behind ; its dorsal surface is much compressed,
with a short, high, triangular crest, whilst on the visceral surface a lower crest runs for
about two thirds of its length and bifurcates behind (Plate XII. fig. A). The *seventh* (biconcave) vertebra (Plate XIII. fig. A) is distinguished by the high crest on its dorsal and
visceral surfaces ; in the middle of the vertebra the neural crest is split into two
branches, diverging in the direction of the posterior zygapophyses and leaving a deep
triangular recess (*a*) between them. The point of divergence forms a kind of summit (*b*)
to this vertebra, which, however, is overtopped by the extremities of the anterior zygapophyses. The place on the neural arch inwards of and behind each anterior zygapophysis is not hollowed out, as in the Galapagos and Rodriguez species, and is even
much less concave than the corresponding portion of the preceding vertebræ. The
anterior and posterior glenoid cavities are completely divided into a right and a left half
in the adult, and less completely in the young specimen. The *eighth* vertebra (Plate XIV.
fig. A) has a bipartite anterior and single posterior condyle, the posterior zygapophyses
being curved downwards to the level of the condyle.

The measurements of the second to seventh cervical vertebræ are as follows :—

		2nd. millim.	3rd. millim.	4th. millim.	5th. millim.	6th. millim.	7th. millim.
Length of centrum	Adult	70	106	105	112	120	105
	Young	37	61	57	64	60	60
Depth of centrum in the middle	Adult	45	48	50	53	68	78
	Young	29	28	29	33	46	46
Horizontal width of middle of centrum	Adult	25	25	27	29	53	56
	Young	17	16	17	18	25	30
Width of anterior condyle	Adult	28	28				
	Young	15	16	...	...	...	...
Width of anterior glenoid cavity	Adult			35	42	50	58
	Young		...	24	27	32	36
Width of posterior condyle	Adult		31	35	44	56	
	Young	...	20	23	28	35	...
Width of posterior glenoid cavity	Adult	27					68
	Young	16	...	...	...	...	39
Distance of outer margins of anterior zyga-pophyses	Adult	44	60	55	59	65	56
	Young	22	34	36	37	40	35
Distance of outer margins of posterior zyga-pophyses	Adult	52	50	48	52	48	65
	Young	29	32	35	36	32	45

A comparison of this table of measurements with those given of the Galapagos Tortoises shows that in the much shorter vertebræ of the Aldabra species all the other dimensions of width, depth, or expansion of articular surfaces and processes equal or even exceed those of the Flat-headed group.

Dorsal vertebræ.—The two heads of the first rib are rather broad, much divergent, leaving a triangular space between them and the first dorsal vertebra. The iliac bones abut against the pleurapophyses of the 9th, 10th, 11th, and 12th vertebræ (counting from the first dorsal vertebra). Their distal extremities participate more or less in the formation of the protuberance for the articulation of the ilium. I am enabled to give the measurements of the dorsal vertebræ of two individuals, viz. of specimens *l* and *m* of the list given above.

Length of centrum of dorsal vertebræ :—

Individual. No.	Length of carapace.	1st. mm.	2nd. mm.	3rd. mm.	4th. mm.	5th. mm.	6th. mm.	7th. mm.	8th. mm.	9th. mm.	10th. mm.	11th. mm.	12th. mm.
l	31 in.	45	70	82	86	80	65	48	39	24	19	16	16
m	29½ in.	43	70	75	76	74	57	40	29	19	17	[lost]	

The number of *caudal* vertebræ is twenty-five, of which the last seven are coalesced into a single bone. The centrum of the *first* (Plate XV. fig. A) is without hæmal crest, and so exceedingly depressed, thin, and flat, that the anterior glenoid surface is reduced to a pair of small lateral tubercles, and the posterior condyle is transversely linear; on the other hand, the neural arch forms an unusually thick mass of an oblongo-tetrahedral shape, bridging over the spinal canal, which in this part is very wide, broader than high.

The præzygapophyses face towards the dorsal, the postzygapophyses towards the visceral side; both are oblique, the former separated from each other by a very narrow groove, the latter by a crest. The continuity of this vertebra to the lumbar is indicated by the development of a long diapophysis flattened and dilated at the base. The *second* caudal resembles the first, approaching in form the third, the principal difference from the first being the reduction in length of the diapophyses. Whilst still retaining the massive neural arch, the *third, fourth*, and *fifth* caudals (Plate XV. fig. B, the fourth) have also their centrum thicker than the preceding vertebræ, the posterior condyle being subglobular, with an obtuse hæmal ridge arising from it. Their præzygapophyses are subvertical, separated by a groove, which is now wide enough for the passage of the blood-vessels, which branch off from those of the spinal canal, and carry the blood through a foramen situated in the median line between every pair of vertebræ to or from the back of the tail. The postzygapophyses are likewise separated by a groove forming one half of the canal for blood-vessels. Diapophyses obliquely directed forwards. In the *sixth* (Plate XV. fig. C) the centrum is shortened, thicker, more distinctly compressed; diapophyses long, directed forward; præzygapophyses obliquely facing dorsad, widely separated by a semicircular notch, and much more expanded than the postzygapophyses. The spinal canal has become narrower, and is deeper than wide. The following vertebræ, from the *seventh* (Plate XIII. figs. D, D′, D″, ninth vert.) to the *eighteenth* (Plate XIV. figs. C, C′, C″, C‴), are built upon the same plan, the gradual modifications being the following:—the centrum becomes shorter and, like the whole vertebra, more depressed; a hæmal ridge soon becomes quite obsolete; the anterior glenoid cavity lower, shallower, and relatively broader; the posterior condyle less prominent, loses its globular form, and becomes lower and transverse; the diapophyses stand at a right angle to the axis of the centrum, lose in length, retaining their relative breadth at the base; the neural arch loses in substance and is depressed, nearly flat above; the neural canal, again, becomes wider than deep. The last seven vertebræ are quite rudimentary and coalesced into a single bone.

Possibly the extraordinary development of the neural arch of the anterior and of the neural arch and centra of the succeeding vertebræ is peculiar to males of such an advanced age as that was from which the foregoing description is taken. In individuals of that sex the tail plays a very important part as an external prehensile, or rather steadying organ, which also differs externally from that of the female in its greater length and by being provided with a large terminal claw. Nearly always the animal carries it bent sidewards under the carapace, generally towards the left side; and therefore I anticipated to find a want of symmetry in some portion of the root of the tail; however, nothing of the kind can be observed. The articulations between the first four vertebræ are so complex that their mobility must be very limited indeed, whilst in the six following the anterior glenoid cavity is disproportionately large for the small size of the condyle; and this seems to be the part where the lateral flexion of the tail is effected.

The length of the entire caudal portion of the vertebral column is 27 inches (that of the carapace being 49 inches).

Limb-bones.—The *scapulary* is stout and massive. The angle at which the scapula proper and the acromium meet is about 100°. The body of the scapula is compressed, trihedral in form, a transverse section through its middle resembling much that of *T. elephantopus*, delineated in the description of that species : its anterior side is convex, as in that species; but neither of the other two surfaces is so decidedly concave. Shaft of the acromium compressed in its proximal half, subtrihedral in its distal portion. The *coracoid* is never ankylosed to the scapula, not even in our largest specimens; its neck is much constricted, compressed transversely to the dilated portion of the bone. The measurements of three examples of this bone are the following :—

	Spec. c. millim.	Spec. l. millim.	Spec. n. millim.
Length of scapula (measured from the suture with the coracoid)...	265	172	200
Circumference in its middle	145	73	92
Longitudinal diameter of glenoid cavity	90	53	
Length of coracoid	150	100	
Greatest width of coracoid	133	80	
Length of acromium	110	78	90

The shaft of the *humerus* (Plate XVI., taken from specimen *c*) is broad and stout, slightly bent, trihedral in the adult, and compressed in the direction from the front backwards. There exists a distinct impression on the outer side of the bone, immediately below the head and ulnar tuberosity, and another deeper one on the hinder side above the trochlea. Both the radial and ulnar edges are much curved. The ulnar tuberosity projects high above the head, which, again, is raised (but not entirely) above the level of the summit of the radial tuberosity. The canal for the blood-vessels on the radial edge of the bone, close to the elbow-joint, is perfectly closed, perforating the substance of the bone from the front to the hinder side.

	Spec. c. millim.	Spec. l. millim.	Spec. n. millim.
Length of the humerus (measured in a straight line from the summit of the head to the middle of the trochlea)	308	180	195
Circumference of the narrowest part of the shaft	160	81	100
Longest diameter of the head	80	39	39
Shortest diameter of the head	60	34	35
Extreme breadth between the condyles	125	66	76

The bones of the *forearm* need scarcely any description beyond giving the measurements. The shaft of the radius has the ulnar edge rounded off and obtuse. In the larger examples the olecranon is scarcely more prominent than in the younger one, and much less so than in *T. daudinii*.

	Spec. c. millim.	Spec. l. millim.	Spec. k. millim.
Length of ulna	105	110	142
Least width of ulna	31	18	22
Length of radius	175	97	130
Least circumference of radius	85	40	63

The general arrangement of the *carpal* bones does not differ from that of the Galapagos Tortoises, in the description of which it will be more fully noticed. The *scaphoid* and *intermedium* are entirely coalesced, even in the younger example; but all the other ossicles are separate.

Important and striking as the differences are that obtain in the *pelvis* of the Flat- and Round-headed Tortoises, the various species of the latter form (as far as known at present) differ but slightly among themselves in this portion of the skeleton. The symphysial bridge is strongly compressed, and raised into a trenchant vertical ridge below and sometimes above. Singularly, in our large specimen *c* all the sutures are as persistent as in the smaller one, so that the bones readily separate; in the latter specimen all the epiphyses are much less ossified, and the protuberances and processes in a much less developed condition, than is observed in other parts of the skeleton of this individual : the pelvis appears to be the last part of the skeleton assuming the proportions and shape of the adult stage.

The horizontal diameter of the pelvis (Plate XVII.) exceeds in length the longitudinal. The iliac bones are broad and stout; the anterior portion of the pubic bones is nearly horizontal, rather obtuse in front, without median crest below; it emits laterally a long strong process, which is obliquely directed outwards. The posterior part of the ossa ischii is of great width, scarcely concave above, terminating in a short hamate posterior process on each side, and provided with an exceedingly high symphysial crest below, which, expanding hindwards, forms a large triangular tuberosity. Lateral margin of the ossa ischii excised in the shape of a C.   Obturator foramina very wide.

	Spec. c. millim.
Longest inner vertical diameter of pelvis (from summit of ilium to symphysis)	200
Longest inner horizontal diameter	230
Shortest inner horizontal diameter (between ilio-pubic prominences)	140
Longest diameter of foramen obturatorium	73
Width of symphysial bridge	35
Depth of symphysial bridge	60
Least breadth of posterior portion of ossa ischii	114
Length of os ilii	183
Least breadth of os ilii	50

The shaft of the *femur* (Plate XIX.) is rather stout, nearly straight, irregularly subtetrahedral, narrower in front than behind. The head has an abbreviated ovate form, and rise to the level of the summits of the trochanters, from which it is separated by a broad and deep cavity. The larger trochanter (a) rises to a higher level than the lesser (b), from which it is separated by a shallow incision. A deeper but narrow depression in the outer margins marks the boundary between the trochanters and the head. The lateral portions of the body of the bone below the trochanters show scarcely any impression. The length of the femur of our largest example (c) is 250 millims., with a least circumference of 128 millims.; the width of the condyles is 120 millims.

Of the *lower leg* and *foot* no part deserves to be mentioned particularly. The *astragalus* and *calcaneum* appear to coalesce at an early stage of growth. And it would appear that in almost all the Round-headed Tortoises the *fibula* is nearly equal to or even exceeds the *tibia* in length (see table of measurements, p. 30), whilst in the Galapagos Tortoises the *tibia* is considerably longer than the *fibula*.

2. TESTUDO DAUDINII.

This name was given by Duméril and Bibron (Erpét. Gén. ii. p. 123) to a skeleton (with carapace) in the Paris Museum which seems to agree closely with the example in the British Museum; it surpasses the latter in size, being 99½ centims. (or 39½ in.) long, and is stated to have come from the " East Indies ;" probably it came on board a vessel returning from the Indian Ocean.

This species is distinguished by the very peculiar form of its carapace, and in this respect as well characterized as *Testudo ephippium* from the Galapagos Islands. I have before me two examples, of which one is an adult male (Plate V.), stuffed, in the British Museum, the carapace being 34½ in. long. The second is a half-grown stuffed male, with a carapace 23 in. long, and belongs to the Free Public Museum, Liverpool. Fortunately the skulls of both specimens could be extracted ; but the remainder of the skeleton is lost. Nothing whatever is known of the history of these specimens; and as no example has reached Europe for a long period, this species may be regarded as probably extinct.

The carapace (Plate V.) is narrow, oblong, and rather deep, not much broader behind than in front; its upper profile from the nuchal plate to the fourth vertebral is nearly horizontalal, though slightly undulating ; but there is scarcely any downward inclination of the first vertebral plate towards the nuchal. A second very characteristic point is the strong reversion of the two anterior and the two posterior marginal plates.

The caudal plate does not participate in the outward reversion of the marginals, but is rather bent inwards, so as to present a convex surface. The anterior and posterior margins are irregularly scalloped. The vertebral plates, as well as the costals, are perfectly smooth and polished, no trace of an areolar surface being visible ; but those of their margins which are in contact with the marginal plates are coarsely striated, as are also the marginal plates themselves. The nuchal plate is of considerable size, broader than long. Caudal plate divided into two unequal portions by an irregular eccentric

F

suture, which is evidently accidental in this individual. However, it is singular that also Duméril and Bibron describe this plate as divided. The *sternum* (Plate IV. fig. B) is deeply concave, broad, with the anterior portion somewhat contracted, truncated in front and behind, the substance of the caudal plates and of the lateral portion of the abdominals being much thickened and very callous. The scutellation of the head and legs does not differ from that of the other Aldabra species, the upper part of the snout being protected by the same large plates, and a longitudinal series of large scutes running along the outer edge of the forearm. The tail is long (about 9 inches), terminating in a long, strong, and flat claw. Duméril and Bibron describe the tail of the skeleton in the Paris Museum as "inonguiculée." The colour is a uniform deep black.

The *younger* specimen differs in some respects, particularly as regards the form of the carapace; but the sculpture of its scutes is exactly such as we should expect in an immature individual of this species, and the reversion of the anterior and posterior marginals is already distinctly indicated. The upper profile, commencing from a protuberance near the hind margin of the first vertebral plate and continued to the extremity of the caudal, is convex; that between that tuberosity and the nuchal plate gently slopes downwards towards the latter. The whole carapace is of a regularly elliptical form, not broader behind than in front, and rather deep. The anterior as well as the posterior margins are indistinctly scalloped; all the upper plates consist of a perfectly smooth areolar portion, which is not much smaller than the scute itself, the striated marginal portion being very narrow, with the striæ shallow and broad. Nuchal plate, as in the adult specimen, large, broader than long; caudal without any division. Sternum slightly concave, with a very shallow notch in front and behind. Tail with a well-developed terminal claw. *Colour* less deep black than in the adult specimen.

The measurements of these two specimens are the following :—

	Length of carapace.		Width of carapace.		Depth of	Sternum.		Caudal plate.	
	In str. line.	Over curve.	In str. line.	Over curve.	carapace.	Length.	Width.	Length.	Width.
Specimen.	in.	in.	in.	in.	in.	in.	in.	in.	in.
Ad. ♂ .	34½	40	19	38	16	23	17½	4	7
Yg. ♂ .	23	29	14½	28	10½	18	12½	1¾	5

The *skull* * of the larger of these examples measures in length from the occipital condyle to the extremity of the snout 4½ inches, or from the extremity of the occipital crest to that of the snout 5 inches, its greatest width being 3¾ inches. It is very similar to that of *T. elephantina*, but differs from it in the following points :—3. The tympanic case, with the mastoid, is much produced backwards, the hind margin of the paroccipital forming a strong curve. 5. The tuberosity on the front margin of the temporal fossa is distinctly indicated. 6. The tympanic cavity is large and deep, the outer tympanic rim having a subsemicircular outline, the convexity being anterior.

* Professor A. Milne-Edwards informs me that the skull attached to the skeleton in the Paris Museum evidently belongs to another animal.

Very few parts of the remainder of the skeleton have been preserved and extracted from the stuffed example. The *ulna* has a length of 150 millims. and a width of 29 millims. in its narrowest part; the *radius* a length of 145 millims. and a circumference of 67 millims., also measured in its narrowest part. Olecranon very prominent, the articular surface forming a steep decline towards that of the radius. Shaft of the radius subcylindrical in its middle. The arrangement of the *carpal* bones does not differ from that of *T. elephantina*. The *tibia* is 137, the *fibula* 139 millims. long.

3. TESTUDO PONDEROSA.

Of this form I have seen one specimen only, the complete skeleton of a female in the British Museum (Plate VI.). It was bought some fifteen years ago by myself for the Zoological Society, in whose gardens it lived for a short time; nothing could be ascertained at the time of its earlier history.

A pelvis of a larger example in the British Museum is referred to this species on account of its complete resemblance to that of the first specimen.

The *carapace* (Plate VI.) is very regularly shaped, a little wider behind than in front, with neither the anterior nor posterior margins reverted, and with shallow notches between the marginal plates. It is rather high, with the sides nearly perpendicular, and with the centre of each of the vertebral scutes raised. Nuchal excision very narrow. Fore part of the shell declivous from the hind part of the first vertebral scute. Caudal scute not reverted, but rather convex. Vertebral and sternal scutes almost entirely smooth, costals and marginals with broad flat concentric striation. Nuchal plate narrow, linear.

·Sternum truncated in front, with a rectangular excision behind; the lateral edges of its front lobe are conspicuously convex along the whole extent of the postgular scutes.

Carapace of a uniform dark horn-colour, nearly black.

The bony substance of the carapace is of considerable thickness in every part, three fourths of an inch in a longitudinal section of the lateral portion of the sternum. The measurements of this specimen are as follows:—

Length of carapace.		Width of carapace.		Depth of carapace.	Sternum.		Caudal plate.	
In str. line.	Over curve.	In str. line.	Over curve.		Length.	Width.	Length.	Width.
in.	in.	in.	in.	in.	in.	in.	in.	in.
27½	36½	19	37	14¼	22½	17¼	2⅜	6½

Osteology.—The *skull* of this species (Plates VIII. & IX.) differs more from those of the preceding two species than these do from each other. It is particularly distinguished by its relatively inconsiderable breadth and the shortness of the occipital crest. 1. The frontal region is very convex and broad, its greatest width (in front of the postfrontals) being more than one half of the distance of the tympanic condyles (outside measure). 2. The upper surface of the parietals is a flat triangle, the outer edges of which are dilated, converge towards, and finally are merged into, the short tetrahedral occipital crest. 3. The tympanic case (with the mastoid *) is not produced backwards; so that the posterior margin of the paroccipital is but little excised. 4. The groove on the lower surface of the occipital, in front of the condyle, is very shallow. A styloid process (Plate IX. fig. B, *a*) projects obliquely outwards on each side of the occipital condyle, and appears to be quite peculiar to this species. 5 to 11. In these points the skull of the present species agrees completely with that of *T. elephantina.* 12. The palate is deeper than in any of the other species, with a very low median crest along its bottom. The distance between the foramina palatina is conspicuously less than that between a foramen and the anterior extremity of the vomer. 13. Anterior surface of the tympanic pedicle very slightly concave. 14. Lower jaw as in *T. elephantina* and *T. daudinii.*

Cervical vertebræ.—On comparing the neck-vertebræ of *T. ponderosa* with those of *T. elephantina* we find the greatest similarity between them. All the distinctive characters by which the latter differs from the Galapagos species are repeated here; and the first and seventh are the only vertebræ which exhibit peculiarities indicative of specific distinctness. In the *atlas* (Plate XIII. fig. C) the lateral portion of the neural arch is very broad, much broader than the zygapophysis, which is nearly as long as that part of the bone which forms the roof of the arch. The centrum (odontoid process) is longer than broad, a tectiform body with the visceral surface compressed and with a broad articular surface in front and behind. The *fourth* is the first vertebra of the series which shows a deep hollow in the neural arch behind the anterior zygapophysis to receive the end of the zygapophysis of the preceding vertebra. The dorsal surface of the *sixth* has a very low median ridge, but no crest ; and in the *seventh* the " summit " is raised quite to the level of the top of the anterior zygapophyses.

Measurements of cervical vertebræ :—

	2nd.	3rd.	4th.	5th.	6th.	7th.
	millim.	millim.	millim.	millim.	millim.	millim.
Length of centrum	36	53	51	63	66	57
Depth of centrum in the middle	27	26	24	26	33	43
Horizontal width of middle of centrum	12	13	13	15	22	24
Width of anterior condyle	12	16	...	...	...	...
Width of anterior glenoid cavity		...	22	22	28	30
Width of posterior condyle	...	19	20	24	28	...
Width of posterior glenoid cavity	15	...	...	...	...	35
Distance of outer margins of anterior zygapophyses	24	31	31	33	34	32
Distance of outer margins of posterior zygapophyses ...	27	29	30	30	30	42

Dorsal vertebræ.—Also this part of the vertebral column differs but slightly from that of *T. elephantina*, except in the singular shortening of the eighth vertebra, which is coalesced with the two following, the sutures between them having nearly entirely

* This bone has been lost by the preparator.

disappeared. For comparison with that species I give the length of the centra of the several dorsal vertebræ :—

Dorsal vertebræ	1st.	2nd.	3rd.	4th.	5th.	6th.	7th.	8th.	9th.	10th.	11th.	12th.
Length of carapace.	mm.	mm.	mm.	mm.	mm.	mm.	mm.	mm.	mm.	mm.	mm.	mm.
T. elephantina...	31 in.	45	70	82	86	80	65	48	39	24	19	16	16
T. elephantina...	29½ in.	43	70	75	76	74	57	40	29	19	17	[lost]	
T. ponderosa ...	27½ in.	35	68	70	70	67	55	42	17	18	17	14	16

The number of *caudal vertebræ* is twenty-five, as in *T. elephantina*, a close and singular approximation to the numbers observed in the Galapagos Tortoises. Throughout this portion of the vertebral column the neural arch is much less thick and high than in *T. elephantina*. The first four vertebræ are much more depressed than in that species and have much longer diapophyses. The transition from the transversely elongate condyle of the fourth vertebra to the narrow and globular one of the fifth is abrupt. The sixth and seventh have their diapophyses obliquely directed forwards, whilst these processes are vertical to the axis of the vertebra from the eighth backwards, and gradually diminish in length. There is scarcely a trace of a hæmal ridge on the first nine vertebræ, the neural surface being longitudinally hollowed out. The last seven are ankylosed as in *T. elephantina*. The length of the entire caudal portion of the vertebral column is 11½ inches (that of the carapace being 27½); but the shortness of the tail is evidently due to sex.

Limb-bones.—The *scapulary* (Plate XXIV. fig. A) is stout and massive. The angle at which the scapula proper and the acromion meet is very obtuse, about 130°. The body of the scapula is compressed, trihedral in form, a transverse section through its middle resembling exactly that of *T. elephantopus* (see p. 67) ; not only is the anterior surface convex, as in that species, but also one of the lateral surfaces is deeply scooped out. Acromion compressed, twisted round its longitudinal axis. The *coracoid* is ankylosed to the scapula, with the neck constricted and slightly twisted. The measurements of this bone, compared with those given of *T. elephantina*, are the following :—

	T. elephantina.		T. ponderosa.
	Spec. c.	Spec. l.	
	millim.	millim.	millim.
Length of scapula (measured from the suture with the coracoid)...	265	172	160
Circumference in its middle	145	73	71
Longitudinal diameter of glenoid cavity	90	53	42
Length of coracoid	150	100	84
Greatest width of coracoid	133	80	80
Length of acromium	110	78	84

The *humerus* resembles so much that of *T. elephantina* that it will suffice to give the measurements :—

	millim.
Length of humerus	180
Circumference of the narrowest part of the shaft	83
Longest diameter of the head	36
Shortest diameter of the head	30
Extreme breadth between the condyles	63

The bones of the *forearm* differ in several respects from those of *T. elephantina* and *T. daudinii*. The shaft of the *radius* is laterally strongly compressed and much narrower than deep. The *ulna* has its radial margin much curved, so that a wide oval vacuity exists between this bone and the radius. The *olecranon* is as prominent as in *T. daudinii*, the articular surface forming a steep decline towards that of the radius. The arrangement of the carpal bones is the same as in the other two species. The comparative measurements of the bones of the forearm in the three species are :—

	T. elephantina.			*T. daudinii.*	*T. ponderosa.*
	Spec. e.	Spec. l.	Spec. k.		
	millim.	millim.	millim.	millim.	millim.
Length of ulna	105	110	142	150	112
Least width of ulna	31	18	22	29	18
Length of radius	175	97	130	145	112
Least circumference of radius	85	40	63	67	46

Of the *pelvis* (Plate XVIII.) two specimens are in the British Museum. All the sutures have disappeared ; the iliac bones are broad and stout, and the vertical and horizontal diameters of the pelvis are nearly equal. The lower part of the pubic bones is gently inclined downwards, prolonged into a tapering lamelliform extremity, and provided below with a sharp median crest. It emits laterally a long, strong process, which is obliquely directed outwards. The posterior part of the ossa ischii is of moderate width, concave above, terminating in a short hamate posterior process on each side, and provided with an exceedingly high symphysial crest below, which, expanding towards behind, forms a large triangular tuberosity. Lateral margin of the ossa ischii excised in the shape of a C. Obturator foramina very wide.

	1st spec. millim.	2nd spec. millim.
Longest inner vertical diameter of pelvis (from summit of ilium to symphysis)	158	116
Longest inner horizontal diameter of pelvis	164	129
Shortest inner horizontal diameter (between ilio-pubic prominences)	97	82
Longest diameter of foramen obturatorium	54	41
Width of symphysial bridge	19	10
Depth of symphysial bridge	45	30
Least breadth of posterior portion of ossa ischii	63	55
Length of os ilii	116	100
Least breadth of os ilii	39	27

The *femur* differs in some measure from that of *T. elephantina*. Its shaft is much flattened in the direction from the front backwards; a narrow and deep cavity separates the head from the trochanters; the two trochanters are entirely separated from each other; and a deep incision forms the marginal boundary between the head and each of the trochanters. The lateral portions of the body of the bone below the trochanters are conspicuously impressed.

The bones of the *lower leg* and *carpus* do not show any noteworthy peculiarity, except that the proximal portion of the *tibia* is deeply grooved on its hinder surface.

	T. ponderosa.	*T. elephantina.*				*T. daudinii.*
		Spec. *c.*	Spec. *l.*	Spec. *a.*	Spec. *P.*	
	millim.	millim.	millim.	millim.	millim.	millim.
Length of the femur	148	250	145	185	...	...
Least circumference of the femur	67	128	76	100	...	...
Width of the condyles	60	120	66	85	...	...
Length of the tibia	113	170	110		136	137
Least circumference of the tibia	48	95	52		70	75
Length of the fibula	108	170	110		131	130
Least circumference of the fibula	34	70	37		51	53

4. TESTUDO HOLOLISSA.

This Tortoise may be considered to be the form which is analogous to the *T. microphyes* of the Galapagos group. Like that species it is perfectly smooth, even from a comparatively early stage of growth. I have seen two carapaces of males, one being fully adult, the other about half-grown: both are preserved in the Museum of the Royal College of Surgeons (Nos. 1020 and 1021)[*].

With regard to their history, we learn from the Cat. Osteol. Ser. Coll. Surg. i. p. 198, that the larger (No. 1021) "was a native of the Seychelle Islands, and was being sent to General De Caen, Governor of the Isle of France, in the French corvette 'Gobemouche,' which was captured by Captain Corbet, of H.M.S. 'Nereida,' and the animal brought to the Cape of Good Hope. It was sent to England by Admiral Bertie, who commanded at the Cape, and remained in a living state at Petworth, the seat of the Earl of Egremont, from August 1800 until April 1810. Its weight was 207 pounds." We have no evidence that any of these Tortoises were indigenous in the Seychelle Islands; on the contrary, it is the belief of the persons best acquainted with this group, that all the individuals kept there in a state of domestication were originally imported from the Aldabra group; and therefore this specimen had probably the same origin as the living female in the Zoological Gardens (see 'Nature,' vol. xii. 1875, p. 260), which evidently belongs to the same species.

[*] I am indebted to Professor Flower, who gave me every facility for a careful and repeated examination of these specimens.

Nothing is known of the history of the half-grown male, beyond that it was a donation from Sir Joseph Banks in 1810.

The *carapace* (Plate VII.) is very regularly shaped, scarcely wider behind than in front, with the anterior and posterior margins slightly reverted, and with very shallow notches between the marginal plates. It is rather high, with the sides nearly perpendicular, and with the centre of each of the vertebral plates slightly raised. Nuchal excision very shallow. Fore part of the shell declivous from the centre of the second dorsal plate. Caudal scute not reverted, but rather convex. The plates of the back, as well as the sternum, are perfectly smooth, without a trace of concentric striæ. Nuchal plate narrow in its upper, and much dilated in its lower portion.

The *sternum* has no excision, either in front or behind. The lateral edges of its front lobe are somewhat emarginate, not convex. As in all adult males, the sternum is deeply hollowed out and provided with thick callosities on the sides and behind. This is already conspicuous in the young male.

Both carapaces are entirely of a uniform deep black colour.

The bony substance of the shell is not particularly thick.

I subjoin the measurements of these three specimens :—

	Length of carapace.		Width of carapace.		Depth of carapace.	Sternum.		Caudal plate.	
	In str.line.	Over curve.	In str.line.	Over curve.		Length.	Width.	Length.	Width.
Spec.	in.	in.	in.	in.	in.	in.	in.	in.	in.
Ad. ♂ (1021).	36	47½	22	48	16	25½	21	3	9
H.-gr. ♂ (1020).	23	28½	14	28	11	17	13	1¾	5¼
Ad. ♀ (living).	32	39½	21	42¼	16¼	25	20½	3½	7

THE EXTINCT RACES OF THE MASCARENES.

In the introductory part of my paper in Phil. Trans. 1875, p. 253, it has been stated that the extinct races of the Mascarenes have a flat cranium, truncated beak, a broad bridge between the obturator foramina, and (it might have been added) the fourth cervical vertebra biconvex. They were thus sharply and structurally differentiated from the Tortoises of the Aldabra group; but there was no common character apparent by which they could be distinguished from the Tortoises of the Galapagos. Such a character can now be pointed out since we obtained more or less complete carapaces from Mauritius and Rodriguez: it consists in the apparently insignificant absence of the suture which divides, in most Land-Tortoises, the gular plate of the sternum into two longitudinal halves. I have besides been able to ascertain that the nuchal plate is invariably absent in all Mascarene Tortoises. Thus, then, what for many years past and at the beginning of these researches seemed to be an almost hopeless task, and what, without the recent explorations instituted in Mauritius and Rodriguez, would have remained an insoluble problem, has found a most simple solution, and there will be in future no difficulty in determining the origin of those carapaces which for generations have been kept in museums, and whose history was veiled in obscurity or forgotten long ago. We may state:—

1. *That the specimens with a nuchal plate (and with double gular) came from Aldabra;*

2. *That the specimens with single gular (and without nuchal) came from the Mascarenes;* and

3. *That the specimens without nuchal and with double gular are Galapagos Tortoises.*

With the aid of this key we can now not only refer various species based by older authors on more or less perfect specimens to their proper geographical division, but also identify them with such as have been distinguished from the osseous remains only.

A. THE GIGANTIC LAND-TORTOISES OF MAURITIUS.

The materials in the British Museum on which the following remarks on the Tortoises of Mauritius are based are the following:—

1. A series of limb-bones and portions of the cranium from the " Mare aux Songes,"

G

chiefly due to transmission by Mr. GEORGE CLARK, C.M.Z.S. *, and obtained by the Trustees of the British Museum in 1865 and following years.

2. The carapace of a Tortoise found at Grand Port, a few years ago, in the same place where the bones of the Dodo were found ("Mare aux Songes"), sent by our esteemed correspondent, L. BOUTON, Esq. [This appears to be the carapace mentioned in Mr. Clark's statement; but no plastron was received with it then or afterwards.]

3. Limb-bones and parts of the pelvis from the district of Flacq, sent by the same gentleman.

4. A perfect pelvis of a very large individual, obtained through C. E. BEWSHER, Esq., of Port Louis.

5. Skulls, sternum, and limb-bones from the "Mare aux Songes," presented by the Hon. E. NEWTON.

Besides, I have had the advantage of consulting the magnificent collection made by the Hon. E. NEWTON, who most kindly submitted it to my examination. It is now in the Museum of the University of Cambridge, and consists of several fragmentary cara-paces, perfect sterna, and a great number of skulls, scapularies, pelves, humeri, and femora. Surpassing in the number of bones the collection in the British Museum, it does not offer the same amount of evidence as regards the variety of races in Mauritius, the objects having been collected at one locality only, the "Mare aux Songes," whilst the remains in the British Museum were obtained at various periods from at least two distant districts of the island.

There can now be no doubt that Tortoises from the Mauritius have been enume-rated in the system under distinct names for many years, although zoologists were ignorant of their origin. Indeed the very first gigantic Land-Tortoise described proves to be a Mauritian species. I allude to the Tortoise described by PERRAULT in the year 1676 (Mém. pour servir à l'Hist. Nat. p. 193, c. tab.) under the name of "La Tortue des Indes," and stated by him to have come from the coast of Coromandel. That no gigantic Land-Tortoise ever came from that coast is certain; but a ship returning thence to Europe would most probably touch at Mauritius. This Tortoise, then, was named by succeeding systematists *Testudo indica*, and finally, by Duméril and Bibron,

---

* The circumstances under which these bones were found will be readily understood from the following abstract of Mr. Clark's "Statement" (Trans. Zool. Soc. vi. p. 51):—"On the estate called 'Plaisance,' about three miles from Mahebourg, in the island of Mauritius, there is a ravine of no great depth or steepness, which apparently once conveyed to the sea the drainage of a considerable extent of circumjacent land, but which has been stopped to seaward, most likely for ages, by an accumulation of sand extending all along the shore. The outlet from this ravine having been thus impeded, a sort of bog has been formed, called 'La Mare aux Songes,' in which is a deposit of alluvium, varying in depth, on account of the inequalities of the bottom, which is formed of large masses of basalt, from 3 to 10 or 12 feet. The proprietor of the estate, a few weeks ago, con-ceived the idea of employing this alluvium as manure; and shortly after, the men began digging in it. When they had got to a depth of 3 or 4 feet they found numerous bones of large Tortoises, among which were a carapace and a plastron pretty nearly entire, as also several crania. . . . These were found imbedded in a black vegetable mould, the lighter-coloured specimens being near the springs."

Testudo perraultii. The carapace of the specimen from which Perrault took his original description is still in the Paris Museum; it is that of a male, about 32 inches long. Unfortunately the sternum is lost, so that the French herpetologists could not be aware that this species belongs to that division of the genus *Testudo* which they distinguished by the presence of eleven sternal scutes only. However, quite apart from the consideration of the sternal characters, the detailed description in the 'Erpétologie générale' offers sufficient evidence of the close affinity of this *T. indica* or *perraultii* with our Mauritian species : " la suscandale simple, *très-élargie* ; la dernière de la rangée vertébrale bombée," are two of the most striking features of the carapaces from Mauritius.

A second carapace, with epidermoid scutes, 16 inches long, preserved in the Paris Museum for more than a century, and without known history, was described by SCHWEIGGER as *Testudo tabulata,* var. *africana,* in the year 1812 [*], and recognized by Duméril and Bibron as a distinct species, *Testudo grayi.* This is likewise a Mauritian Tortoise ; and as it has the same undulated vertebral profile as the carapace subsequently described in this work as possibly belonging to *T. triserrata,* these two individuals may eventually have to be referred to the same species.

For the present I must be satisfied with having thus indicated the natural affinity of *Testudo indica* and *T. grayi* with the Mauritian species described from osseous remains ; but a much more complete series of carapaces (and, indeed, of carapaces with their skulls and other parts of the skeleton) will be required before we can venture to decide whether or not those two species are identical with the species distinguished in this work. A carapace with so straight a vertebral profile as that delineated and described of *T. perraultii* is not represented among the specimens collected by Messrs. Bouton and Newton.

The Mauritian Tortoises were inferior in size to those of Aldabra and Rodriguez, the majority of the remains belonging to individuals with carapaces from 2 to 3 feet long ; specimens above this size must have been exceedingly rare.

In making the collections mentioned above, no care was taken to keep separate such of the bones as were found in juxtaposition, and which, therefore, might reasonably have been presumed to belong to the same individual. Hence it would be most difficult (nay, impossible) to avoid errors, if we were to attempt to specifically group the modifications which can be observed in almost all the different kinds of bones ; and it will be the safer plan to adopt in their description an analytical method—that is, to take up the various kinds of bones in their natural succession and to describe the modifications of each. Perhaps, in a few instances, it will be possible to indicate the specific unity of the types thus distinguished [†].

[*] Königsb. Arch. Naturwiss. i. p. 322.

[†] In a preliminary notice published in 'Nature,' 1875, p. 297, before I had access to Mr. Newton's collection, I had not fully realized the difficulty of referring each modification of bone to its proper species ; and, more especially, I fell into the error of referring to *T. boutonii* a humerus which proves to belong to *T. inepta.*

Skull.—There are altogether thirteen skulls in a more or less fragmentary condition, and five mandibles, in the collections of the British and Cambridge Museums. They can be referred to two species—the first having three serrated dental ridges along the lower jaw (*Testudo triserrata*), the second having only two ridges (*Testudo inepta*).

a. *Testudo triserrata* (Plate XXIII. fig. A).—1. Frontal region perfectly flat, broad, its greatest width being a little more than one half of the distance of the tympanic condyles.   2. Only the anterior half of the parietal forms a flat surface, the posterior being compressed into a trenchant crest passing into the very long occipital spine.   3. The tympanic case with the mastoid is produced backwards, so that the paroccipital margin appears as a deep semicircular excision.   4. The impression on the lower surface of the occipital in front of the condyle is very shallow.   5. On the front margin of the temporal fossa, corresponding to the suture between parietal and tympanic, immediately in front of the foramen carotidis externæ, there is a very large and projecting flat condyle-like tuberosity for the insertion of a portion of the temporal muscle.   6. Tympanic cavity large, with a nearly circular outer tympanic rim, which is posteriorly interrupted by the notch for the Eustachian tube.   7. This notch is not very remote from the columellar foramen, which is on an elevated ridge crossing transversely the tympanic cavity.   9. The front margin of the intermaxillaries projects but slightly beyond that of the frontals ; so that the plane of the nasal opening is nearly perpendicular and as broad as high.   10. The inner nostrils are advanced, not very distant from the end of the snout, and, on the palatal view of the skull, hidden for the greater part below the alveolar lamella of the maxillaries.   11. The intermaxillaries are short, about one third the length of the maxillaries ; their foremost portion is deeply hollowed out below, and vertically bent downwards to form the beak.   The suture between the intermaxillary and vomer is immediately behind the inner angle of the alveolar edges of the maxillaries.   The alveolar maxillary surface is provided with three denticulated ridges, the inner of which runs parallel to, but at a distance from, the inner alveolar edge.   12. The palatal region is much less concave than in the Aldabra Tortoises, and without median longitudinal crest.   13. Anterior surface of the tympanic pedicle deeply excavated.   14. Lower jaw with a triple serrated alveolar ridge, the two inner ridges being rather close together and belonging to the same raised osseous tract.

b. *Testudo inepta* (Plate XXIII. fig. B).—The skull of this species is so similar to that of *Testudo triserrata* that only the following important points in which it differs need be mentioned.   11. The alveolar surface of the maxillary shows three denticulated ridges, one being confluent with the outer, and one with the inner alveolar margin. 14. Lower jaw with two serrated ridges only.

To judge from the skulls, both these species must have had smaller heads than the Galapagos Tortoises. The largest of our series of skulls belong to *Testudo triserrata*, one of them being 50 lines long (to the end of the occipital condyle) and 38 lines broad (between the tympanic condyles); the largest skull of *Testudo inepta* measures 46 lines in length and 34 lines in width.

Carapace.—Among the carapaces two very distinct forms may be recognized, which, however, have this in common—that the centre of the last vertebral scute is raised in a very convex hump, which is separated from the preceding vertebral by a deep transverse depression. The shell of both forms is thin and fragile (though less so than in the Rodriguez Tortoise); but towards the margins it is much thickened, especially in the second form, and four or five times as thick as on the convex portions. The skulls belonging to the carapaces not having been found, or having been mixed with others, it remains uncertain to which of the species distinguished from the skulls the carapaces should be referred.

a. *Carapaces possibly belonging to* Testudo triserrata (Plate XX. fig. D).—The greater portion of the upper part of a carapace (32 in. long and about 20 in. broad) and the hinder third and the foremost part (including first vertebral and marginals) and five sterna, of adult males, are in Mr. Newton's collection. This carapace is more depressed than that of any other gigantic Tortoise. The upper profile is much undulated, the second, third, fourth, and fifth vertebrals being raised in the middle, and the humps separated by deep transverse impressions. The first vertebral is flat, its median profile gently sloping downwards towards the front margin. The decline of the upper profile from the hump of the last vertebral to the hind margin of the caudal is steep and straight. Also the costal plates participate in a slight degree in the peculiar formation of the vertebrals.

Most of the marginal plates are lost: the first barely touches the antero-inferior angle of the first costal; the last two are as high as broad; and the caudal is much broader than high.

The margins of the shell are not visibly indented; but indentation may have disappeared with the epidermoid scutes.

None of the sterna collected by Mr. Newton can be fitted to the individual carapaces or fragments of carapaces. They are those of adult males, deeply concave, and callous on the sides. Two of the largest are 18 and 19½ in. long, and respectively 14½ and 15 in. broad. The lobes (Plate XX. fig. E) are short, especially the posterior, which is about twice as broad as long and broadly truncated behind. The single gular plate is triangular, with rounded front margin and acute posterior angle, wedged in between the postgulars. Pectorals extremely narrow, almost linear. Abdominals very large, their length being two fifths of that of the entire sternum.

b. *Carapaces possibly belonging to* Testudo inepta.—A nearly entire specimen (Plate XX. figs. A–C), but without plastron, has been sent by M. Bouton; it was discovered in 1865, in the "Mare aux Songes." A large fragment, representing the posterior third of the carapace, and perfectly agreeing with the other specimen, is in Mr. Newton's collection. This form of carapace is much higher and narrower than that of *T. tri-*

serrata (?). Our specimen has a length of 22 inches and a greatest depth of 8½ inches. The thickness of the shell scarcely exceeds 2·5 millims. along the central portions, but gradually increases towards the margins. It is that of a fully adult animal, as all the sutures have disappeared ; nevertheless we cannot draw a conclusion from this single individual as regards the maximum size to which this species attained, because in most forms of Testudinata we find examples of smaller and larger size, independently of age.

The general shape of the carapace is very peculiar, and very different from any of the other gigantic Tortoises. The anterior three fifths of the shell are very convex and much bulged out in every direction ; the posterior two fifths slope towards the much-expanded hind margins ; but this slope is rendered very uneven by the fourth and fifth vertebral plates being much raised in the middle, and the two elevations being again separated from each other by a depression. The margins of the shell also of this species are not conspicuously indented. The sutures between the first, second, third, and fourth vertebrals are of nearly the same length (3″ 11‴, 4″ 9‴, and 4″ 6‴), whilst the suture between the fourth and fifth vertebrals is only one half that length (2″ 3‴). The first marginal plate is entirely limited to the anterior vertebral, without extending to the adjoining costal. The undivided caudal gently slopes backwards and downwards, is much expanded, not turned in or reverted behind ; its hind margin is strongly curved. It measures 5½ in. in its greatest width, and 3 in. 2 lines in length.

Vertebral column.—A single example of the fifth vertebra of the cervical series is preserved (Brit. Mus. no. 39930); there is no means of ascertaining whether it belongs to *T. triserrata* or *T. inepta*. As regards form, it is nearly intermediate between those of *T. ponderosa* and *T. vosmaeri*; there is scarcely a trace of a crest on the dorsal surface. The individual may have been somewhat inferior in bulk to those whose vertebræ are used in the subjoined table for comparison.

	T. ponderosa. millim.	T. vosmaeri. millim.	Mauritian. millim.
Length of centrum	65	73	61
Horizontal width of middle of centrum	17	14·5	14
Width of anterior glenoid cavity	28	29	26
Breadth of posterior condyle	30	20·5	22

Scapulary.—Among the numerous specimens three types may be distinctly made out :—

a. The majority (Plate XXIV. fig. B.) have the distal portion of the acromium, as well as the proximal half of the scapula, distinctly trihedral, the anterior side of the latter being convex, so that a transverse section through that part of the scapula represents the following outline (fig. 1). The acromium is straight, the coracoid lamina sometimes deeply grooved. The whole bone is much more slender than in the Aldabra Tortoises, but less so than in the Rodriguez species. Although these specimens differ

Fig. 1.

considerably in size (the scapula proper of the largest individual being $6\frac{1}{2}''$ long, and that of the smallest $4\frac{1}{2}''$) it is remarkable that all have the coracoid firmly ankylosed to the scapula, the suture having entirely disappeared.

This form of scapula, which may be referred to *Testudo triserrata*, does not differ from two other scapularies in the British Museum, obtained together with a femur and fragments of pelvis, for which I have proposed the name of *Testudo leptocnemis*.

b. In the second type, of which I have three specimens (Nos. 39937 and 39938 in the British Museum and one in Mr. Newton's collection) before me, the distal portion of the acromium is compressed and its end is curved ; the proximal half of the scapula is trihedral ; but the anterior and acromial sides have a deep longitudinal impression, so that a transverse section through this part of the scapula represents a figure like the following (fig. 2).   Otherwise the bone is of the same build as in the first type ; but, although these specimens again differ considerably in size (the scapula proper of the larger being $6\frac{1}{4}''$ long, and that of the smaller $4\frac{3}{4}''$), the coracoid never became ankylosed to the scapula, and is lost.   This is a most singular distinction from *T. triserrata*.   The species to which this type of scapula belongs is possibly *Testudo inepta*.

Fig. 2.

c. The third modification, of which I have seen two examples (No. 39936 in the British Museum and one in Mr. Newton's collection), approaches closely to the preceding in having the same compressed acromium, with the end slightly curved. The coracoid, likewise, was not ankylosed to the scapula ; but the scapula itself is strongly compressed, in no part trihedral, and a transverse section would be of the following shape (fig. 3).   This bone is the only one on which our knowledge of a supposed fourth species, *Testudo boutonii*, rests ('Nature,' 1875, p. 207).   But having found a perfectly analogous instance in the scapula of the Rodriguez Tortoise, I must regard this modification as a case of anomaly only.

Fig. 3.

Measurements.

	T. triserrata. N*. millim.	No.39930. millim.	No.39940. millim.	T. leptocnemis. (B. G.) millim.	T. inepta. No.39937. millim.	No.39938. millim.	T. boutonii. millim.
Length of scapula (measured from the suture, with coracoid)	162	140	114	150	161	114	128
Circumference in its middle	72	62	50	59	65	39	53
Longitudinal diameter of glenoid cavity	43	36	32	42	lost		
Length of coracoid	74	61	54 } damaged				
Greatest width of coracoid	61	59	51 }				
Length of acromium	70	54	52	62	75	55	61

* From Mr. Newton's Collection.

Humerus.—The two collections contain together 18 specimens, among which again three types may be distinguished.

a. The first type (Plate **XXV.** fig. A) is represented by six humeri with a slender and strongly bent shaft, and with a deep hollow behind the head of the humerus and between the tuberosities. The head is globular, or, in large specimens, ovoid, and nearly entirely raised above the level of the summit of the radial tuberosity. A groove indicating the course of vessels becomes distinct on the convex side of the distal third of the bone close to the radial edge, and perforates the lower end of the bone very near to the margin of the cartilage of the joint. The slight build of the humeri of this type would seem to indicate that they belong to the species with the thinner carapace, viz. *T. triserrata.*

b. In the second type the hollow behind the head is absent or scarcely indicated. Otherwise it does not differ from the first, and is, perhaps, nothing but an individual variation.

c. The third type (Plate XXV. fig. B), represented by eight specimens, has a much stouter and straighter shaft; it is trihedral nearly in its whole length, a strong ridge generally running down from the tuberosities to the radial margin of the condyle. On account of its greater strength, this form of humerus may be associated with the heavier carapace of *Testudo inepta.*

Measurements.

	T. triserrata.			T. inepta.	
	No.39947.	No.39946.	No.39942.	No.39944.	No.39943.
	millim.	millim.	millim.	millim.	millim.
Length of humerus (measured in a straight line from summit of head to the middle of trochlen)	137	110	184	119	160
Circumference of the narrowest part of shaft ...	55	47	82	63	88
Largest diameter of head	28	24	46	29	39
Shortest diameter of head	25	20	30	20	27
Extreme breadth between condyles	45	38	76	43	58

Ulna.—Two examples of ulna are preserved in the British Museum series: the smaller is that of an animal still growing; the larger (No. 39948) belongs to the left side and is 110 millims. long; it has a comparatively narrow shaft, not more than 15 millims. wide in its narrowest portion. The shaft is considerably twisted round its longitudinal axis, so much so that the transverse diameters of its proximal and distal dilatations stand nearly at right angles to each other. The olecranon is well developed.

Radius.—Four specimens are preserved in the British Museum series, differing considerably from one another, but leaving it quite uncertain whether these differences are merely accidental and individual or specific.

a. One radius of larger size (No. 39950) (Plate **XVIII.** fig. D) and another smaller one (No. 39953), both of the right side, have the shaft rather smooth, subtetrahedral, with a large tuberosity on the posterior surface in the middle of the length. The semilunar facet for the articulation with the humerus is comparatively narrow (40 : 21).

b. A third radius (No. 39952) agrees with the first two in the shape and smoothness of the shaft, which, however, is a little more slender. The tuberosity occupies also the same place; but the semilunar articular facet is conspicuously broader and shorter (37 : 23).

c. The fourth radius (No. 39951) is of the left side of a large individual, and distinguished by rough crests and protuberances, especially in its lower fourth. Its shaft is more compressed, and the tuberosity on its hinder surface placed above the middle and connected by a strong ridge with the lateral facet for the articulation with the ulna. The semilunar articular facet has nearly the same form as in the first type (39 : 21).

Measurements.

	No. 39950. millim.	No. 39953. millim.	No. 39952. millim.	No. 39951. millim.
Length of radius	127	97	125	132
Circumference in its narrowest part	53	32	47	52
Longitudinal diameter of semilunar facet	40	24	37	39
Transverse diameter of the same	21	18	23	21

Pelvis.—The two collections contain nine perfect specimens, and fragments of as many other individuals, referable to three distinct types:—

a. Seven out of the nine perfect specimens belong to a type (Plate XXVI. fig. A and Plate XXVIII. fig. C) which is distinguished by the short vertical and long horizontal diameter, thus affording sufficient ground for associating it with that low carapace which has been referred to *T. triserrata*. The male and female specimens of this type can be clearly recognized by the great difference in size. The ossa ilii are short and broad, far apart, so that the vertical diameter of the pelvis does not much exceed, or may be even shorter than, the horizontal. A strong ridge rises from the middle of the outer surface of the bone and ascends to the anterior spine. The lower part of the pubic bones is strongly inclined downwards, concave above in the male, and slightly convex in the female, and emits laterally a strong process, dilated at its end and more or less obliquely directed outwards. The posterior part of the ossa ischii is of considerable width, sometimes concave above, always with a trenchant symphysial crest beneath; in one female specimen (No. 39935) it develops on each posterior corner a hamate process pointing outwards. The obturator foramina are comparatively narrow, sometimes scarcely wider than the symphysial bridge between them, which is perfectly flat, without median ridge.

In two of the male specimens (both in Mr. Newton's collection) all the ridges and crests are much more prominent, the edges sharper, every impression and hollow deeper

H

than in the other specimens. They evidently belonged to very old individuals; and the atrophied condition of the bone is no doubt due to age. The individual to which one of these pelves belonged is represented also by a humerus showing the same peculiarity. The female pelvis is scarcely half the size of the male.

b. Of the second type (Plate XXVI. fig. B, and Plate XXVIII. fig. B) two perfect male examples are in the British Museum, and two fragments of a female and male in Mr. Newton's collection. It is readily distinguished from the first type by its considerably longer and narrower ossa ilii, the longitudinal diameter of the pelvis much exceeding in length the horizontal. On the outer surface of the ilium scarcely a trace of a ridge is visible in the upper dilated portion. The lower part of the pubic bones is gently inclined downwards and slightly concave above, and emits laterally a long, strong, nearly styliform process, obliquely directed outwards. The posterior part of the ossa ischii is of moderate width, deeply concave above, with a trenchant symphysial crest beneath, which, expanding towards behind, forms a large triangular tuberosity. The upper posterior corner of each side is developed into a smaller tuberosity, pointing upwards. The obturator foramina are of moderate width, considerably wider than the symphysial bridge between them, which is provided with a median ridge above.

This type probably belongs to *T. inepta.*

c. The specimens of the third type are in a fragmentary condition, and were sent to me by M. Bouton with the scapula and femur. They came evidently from a deposit different from that in which the Dodo bones and the other Mauritian Tortoises were found, as they are not stained with the brown peat-colour, nor show any traces of rootlets of plants having been attached to them. Their surface is of a dirty white; and many of the edges are worn, as if the bones had lain in running water between pebbles. The four fragments belong to four individuals, each consisting of an acetabulum with the neighbouring parts; two have nearly the entire ilium preserved, one an obturator foramen with symphysial bridge, all the lateral pubic process. The posterior ischiadic tuberosity and the anterior pubic dilatation are lost.

These pelves resemble in every respect, as far as we can see, the first type; but the os ilii is narrower, though considerably less so than in the second type. The lateral pubic process is slightly thickened at the end.

They belong to the species *Testudo leptocnemis,* which is characterized chiefly by its femur, and of which we have noticed the scapula above.

Measurements.

	T. triserrata.				T. inepta.		T. leptocnemis.		
	N.*	No. 30032.	No. 30034.	No. 30035.	75.10.20.1.	No. 30031.	B 1.	B 2.	B 3.
	♂	♂	♀	♀	♂	♂			
	mm.	mm.	mm.	mm.	mm.	mm.	mm.	mm.	mm.
Longest inner vertical diameter of pelvis (from summit of ilium to symphysis) ...	126	135	95	90	156	140			
Longest inner horizontal diameter of pelvis	128	140	90	84	122	102			
Shortest inner horizontal diameter (between ilio-pubic prominences)	89	82	60	60	94	72			
Longest diameter of foramen obturatorium	35	32	20	19	40	36			21
Width of symphysial bridge	26	27	22	16	30	22			26
Breadth of posterior portion of ossa ischii	64	55	34	38	65	50			
Length of os ilii	112	110	75	73	130	118	120	105	
Least width of os ilii	84	28	22	19	28	21	27	26	

Femur.—Thirteen out of fifteen specimens are perfectly similar; and having been found in the "Mare aux Songes" with the carapaces of *T. triserrata* and *T. inepta*, we may infer that the femurs of these two species do not differ from each other.

a. All these specimens (Plate XXIX. fig. A) are rather stout, with a nearly straight subtetrahedral shaft; its distal portion is much dilated, consequently its outer margin much curved. The ovoid head of the femur is separated from the trochanters by a deep and broad cavity, out of which arises a more or less indistinct smooth groove, dividing the trochanters. The head itself does not rise above the level of the summit of the larger trochanter.

b. The two remaining femurs (Plate XXIX. fig. B) offer chief evidence of the existence of another species, for which I have proposed the name *Testudo leptocnemis.* They were found together with the scapulæ and pelves described above, and differ from the preceding species by the slender shaft and much less dilated condyles. The shaft is less distinctly tetrahedral, the anterior side being somewhat narrowed and very convex. The summits of the ovoid head and larger trochanter are on the same level; and the hollow between head and trochanters is deep and broad; also the groove between the trochanters is distinct.

* Newton collection.

Measurements.

	T. vi	T. vi	T. inepta.	T. leptocnemis.	
	No. 39954. millim.	No. 39958. millim.	No. 39959. millim.	No. 39955. millim.	B7. millim.	B8. millim.
Length of femur	152	133	116	152	135	140
Least circumference of femur	75	71	56	75	63	60
Width of condyles	62	56	45	50	48	50

Tibia.—Two examples of this bone are in the collection of the British Museum. Although they do not much differ in size, the ossification of the apophyses of the smaller one (No 39960) is in a comparatively much more backward stage than in the larger example (No. 39961) (Plate XXVII. fig. D), which is 125 millims. long, with a circumference of 50 millims. in its narrowest part. A still larger specimen, 140 millims. long, is in Mr. Newton's collection. The articular surface is divided into two portions for the reception of the condyles, of which the inner is deeply concave, the outer nearly flat. The shaft is slightly bent. The distal end develops a prominent inner malleolus, separated from the opposite part of the articular surface by a deep hollow.

Fibula.—Only one fibula (No. 39962) (Plate XXVIII. fig. H) is preserved; it is of the right side of a very large individual, evidently the same of which the radius (No. 39951) has been described. It is a straight slender bone, 134 millims. long, nearly cylindrical in the middle, and gradually widening below. Its proximal end presents an oblique concave facet to the articular surface of the outer condyle, its distal end a subquadrangular convex surface to the tarsus. Its lower and upper portions and its hinder surface are provided with rough tubercular crests and protuberances, which is evidently but incidental to the age of the individual.

B. The Gigantic Land-Tortoise of Rodriguez.

The scanty historical evidence preceding the final extinction of the Rodriguez Tortoise has been referred to in the Introduction.

Remains of this Tortoise had been discovered and had reached Europe many years ago; but no particular attention was paid to them. M. J. Desjardins, one of the first explorers of the fauna of Mauritius, sent a bone of a Tortoise, found, in 1786, in a cave in Rodriguez, with some remains of the Solitaire to Paris [*], where they were examined by Cuvier and Blainville, who erroneously stated them to have been recently found under a bed of lava in Mauritius [†]. Another Mauritian naturalist, C. Telfair, in searching, in 1832, for bones of the Solitaire in Rodriguez, succeeded in obtaining " numerous bones of the extremities of one or more large species of Tortoise," which were presented to the Zoological Society of London, and exhibited at one of the meetings [‡]. These bones were still in the possession of the Society three or four years

* Proc. Comm. Zool. Soc. ii. p. 111.   Strickland and Melville, ' The Dodo,' pp. 51–53.
† Edinb. Journ. Nat. Sci. iii. p. 30.   ‡ Proc. Zool. Soc. 1833, p. 31.

before the publication of Strickland and Melville's memoir on the Dodo (1848); but no further attention being paid to them, they were lost. Another portion of TELFAIR's collection was presented by him to the Andersonian Museum at Glasgow, where they are still preserved.

Some well-preserved bones, kindly sent to the writer by M. BOUTON, of Port Louis, in 1872, satisfactorily proved that the Tortoise of Rodriguez is distinguished from all its congeners by well-marked characters (Ann. & Mag. Nat. Hist. 1873, xi. p. 397); but it was only when these remains were supplemented by those preserved in the Andersonian Museum at Glasgow and intrusted to me by the curators of that institution for examination, and when, finally, the extensive series collected during the Transit-of-Venus Expedition arrived, that our knowledge of its specific characters became tolerably complete. No further important additions can be expected from Rodriguez, with the exception of the small bones of the foot and caudal vertebræ; and these will be but of small value, unless they be found in their natural connexion.

With the aid of the carapaces brought home by Mr. SLATER, we are now enabled to recognize the Rodriguez Tortoise in some carapaces which reached Europe in the last century, probably during the lifetime of the species, and which we find noticed by the following herpetologists :—

1. SCHOEPFF (Histor. Testud. 1792, p. 103, pl. 22. fig. B) has reproduced a sketch of a Tortoise 2¾ feet long, which was communicated to him by Vosmaer, who examined the specimen which then was in "Museo Principis Arausionensis" in the Hague. This seems to have been a male, with a carapace very similar in form to that of the male described below; its front and hind margins, being still provided with the epidermoid scutes, have an undulated outline. Schoepff was informed by Vosmaer that the carapace had been brought from the Cape of Good Hope; and expressing himself uncertain whether it should be considered a distinct species, or a sexual, local, or individual variety of the Tortoise described by Perrault, he named it "*Testudo indica Vosmaeri.*"

2. DUMÉRIL and BIBRON recognized Schoepff's Tortoise in a skeleton with complete carapace in the Paris Museum. The description of the specimen, whose shell measured 75 centims. over the curvature, again perfectly agrees with our male specimen, and supplies a detailed account of the outer epidermoid covering. The authors adopt the binomial term, "*Testudo vosmæri,*" which, of course, supersedes that proposed by myself (*Testudo rodericensis*, Ann. & Mag. Nat. Hist. 1873, xi. p. 397). By the singular resemblance of the general form of the male of this species to that of some of the Galapagos Tortoises, they were led into the error of supposing that *T. vosmæri* came from the Galapagos Islands (Erpétol. Génér. ii. p. 140).

3. A second specimen, probably a young female, likewise in the Paris Museum, and without known history, was considered by the French herpetologists a distinct species, *Testudo peltastes* (ibid. p. 138). This description agrees in every respect with our young carapaces from Rodriguez.

The Rodriguez Tortoise differs from the Mauritius and Galapagos Tortoises by the more slender build of all the various parts of its skeleton; its neck must have been capable of still greater flexion, as is evidenced by the deep postapophysial impressions or actual perforations of the cervical vertebræ. Although careful comparative measurements show beyond doubt that this Tortoise had longer limbs and a longer neck than even some of the Galapagos Tortoises, yet, taking also into consideration the extreme thinness and fragility of its carapace, we must infer that this general slenderness of the bones must have been partially due to the same cause, probably a diminished supply of the calcareous salts, or a diminished power of assimilation of them.

The bones collected by the naturalist of the Transit-of-Venus Expedition belonged to several hundred individuals; and there are in some cases as many as forty specimens of one and the same bone in the collection; yet no variation in structure equivalent to that observed among the Galapagos, Aldabra, and Mauritius Tortoises could be detected, so that evidently in this small island there was room for one species only. The only variation which is worth recording is one which can be explained as a sexual difference, the female having been of a smaller size and somewhat stouter form than the male, as is the case in the other Gigantic Tortoises.

Bones far exceeding in size the majority of their kind are not rare, and prove that the Rodriguez Tortoise was quite equal in bulk to *Testudo elephantina*, many (probably male) individuals having had a carapace 4½ feet in length. From Duméril and Bibron's descriptions we learn that the scutes were perfectly smooth or nearly so, and that the shell of the adult was black, whilst the young were of a lighter brown colour, the sternum being dotted with yellow.

Description of the Osseous Remains.

Skull (Plate XXIII. fig. C).—The skulls of twenty-four individuals of various ages, in a more or less imperfect condition, have been collected. They show very little variation beside that which is simply due to age. Two of the largest and most perfect are respectively 4¼ and 4⅕ inches long, measured from the intermaxillary to the occipital condyle, and 3½ and 3 inches broad at the widest part, viz. between the tympanic processes. 1. The frontal region is very broad and flat transversely, sloping in a more or less strong curve towards the nostril, its greatest width (in front of the postfrontals) being as much as, or not much less than, one half of the greatest width of the skull. 2. Only the foremost part of the parietals forms a flat surface, the remainder being compressed into a crest that passes into the long and narrow occipital spine, which rises scarcely above the level of the upper surface of the skull. 3. The tympanic case, with the mastoid, is produced backwards, the hind margin of the paroccipital forming a rather strong curve. 4. There is a shallow impression or hollow in front of the occipital condyle. 5. On the front margin of the temporal fossa, corresponding to the suture between parietal and tympanic, and directly in front of the foramen carotidis externæ, there is a large, prominent, concave, rough tuberosity for the insertion of a portion of the temporal muscle; exteriorly a narrow and deep groove separates this tuberosity from the zygomatic arch; inside of it runs a sharp ridge in an oblique direction from the tympanic condyle to the upper limit of the temporal fossa. This ridge separates the foramina by which the nerves and blood-vessels that leave the cerebral cavity by the foramen sphenoidale (portions of the nervus quintus and carotis cerebralis) issue from the substance of the bone. These external foramina vary somewhat in number and position; but there are generally two behind and a smaller one in front of the ridge. 6. Tympanic cavity large, but somewhat constricted in the interior portion of its entrance; the outer tympanic rim has a rather irregular outline, not well defined in front, and with a deep and rather broad notch in the posterior part of its circumference for the passage of the Eustachian tube. 7. The columella is attached to and rests upon a long, nearly straight and rather sharp ridge, which runs from the notch mentioned to the stapedial foramen. 8. The posterior wall of the inner tympanic cavity, which in fresh examples is formed by cartilage and an open space in the preserved skull after maceration, is but limited in extent, about one seventh of the area of the tympanic opening. 9. The front margin of the intermaxillary projects beyond that of the frontal, the nasal opening being about as wide as long. 10. Although the position of the choanæ is advanced forwards, yet on the palatal view of the skull the greater portion of them may be seen uncovered by the alveolar lamellæ of the maxillaries; the front extremity of the vomer which separates the choanæ is very narrow and slender. 11. The foremost part of the intermaxillary is hollowed out below and vertically bent downwards to form the truncated beak. The suture between the intermaxillary and vomer is immediately behind the inner angle of the alveolar edges of the maxillaries. 12. Palatal region rather deeply concave and broad, a longitudinal median crest being scarcely indicated. The distance between the foramina palatina is much less than that between these foramina and the anterior extremity of the vomer. Outer pterygoid edge rather elevated and sharp. 13. Anterior surface of the tympanic pedicle with a moderately deep impression. 14. Lower jaw with a double alveolar ridge; its symphysial portion simply vertical, without a backward expansion of the lower margin of the bone.

Carapace (Plates XXI. and XXII.).—Tolerably complete carapaces of five individuals were found; unfortunately they are of more or less young age, the two largest being only 16½ and 21 inches long. Four of them are very similar to one another, whilst the fifth and largest, though presenting the same specific characteristics as the others, differs considerably in shape. The carapace is distinguished by its extreme thinness and fragility, the lateral portions being only 1 millim. thick. Although the weather must have had some effect upon the preservation, exposure to it cannot have been the only cause, and can only slightly have deteriorated the surface, as the lines of impression of the epidermoid plates are distinctly preserved. Another characteristic of this species is the extraordinary width of the anterior and posterior openings, the sternum being very short, with both lobes contracted.

The normal shape of the carapace of the young (Plate XXII.) is a regular oval with subvertical sides, a moderate declivity in front, and a strong declivous expansion behind. No part of the margin is visibly reverted ; and the vertebral plates are scarcely raised in their central portion. There is no nuchal plate ; and the gular and broad caudal were undivided.

In two out of the four normal specimens the *sternum* is slightly concave, in the two others nearly flat. The anterior lobe is broader than long, truncated in front, or more or less tapering. The posterior is still shorter, with a very broad subtruncate or somewhat rounded hinder margin.

With regard to the fifth specimen (Plate XXI.), we may assume as almost certain that it was of the male sex, and rather more advanced in age than the others. The front portion of the dorsal shield is unfortunately broken away and lost ; but the perfectly level central profile of the three middle vertebral scutes shows clearly that there was no declivity of the outline in front; and the lateral margins of the front opening appear to have been somewhat reverted. The hind portion of the dorsal carapace is much produced backwards; and its marginal plates are expanded in a quite remarkable manner, being almost horizontal. The anterior portion of the sternum is much constricted, tapering in front, with concave lateral margins. Posterior lobe short and truncated as in the other specimens. The postgular plates approach so close to the abdominals that the intervening portion of the pectorals is reduced to a narrow strip.

The following are the principal measurements of three of the specimens :—

	Length of carapace.		Width of carapace.		Sternum.		Caudal plate.	
	In str. line.	Over curve.	In str. line.	Over curve.	Length.	Width.	Length.	Width.
Spec.	in.	in.	in.	in.	in.	in.	in.	in.
♀	16½	20¼	10	22	11	9	1⅞	4
♀	16	19	9½	20	11	8¾	1½	4
♂	21(?)	24(?)	11½	20½	15	10½	1¾	4¾

Cervical vertebræ.—All the cervical vertebræ are of a comparatively slight build, with thin walls for the spinal canal, with simple and low crests, and with the articular processes but slightly diverging.

Of the majority a considerable number of specimens have been collected ; and we have sufficient means of judging of the amount of variation in each vertebra, also of comparing it with the corresponding bone in the other allied species ; but, unfortunately, we do not possess a perfect series of the vertebræ of one and the same individual ; and although some more or less perfect series may be selected as apparently belonging to the same individual, such connexion between them is by no means a matter of absolute certainty : yet it would have been of some importance to obtain correct data as regards the relative lengths of the several segments of the cervical column.   In nearly all the cervical vertebræ two forms may be distinguished, of which one is still more slender than the other; but, with this exception, there is a great uniformity of structure or want of variation.

No portion of the *atlas* has been preserved. Of the *second* vertebra we have two specimens, both of young individuals. It is provided with a moderately high neural crest, extending along the whole length of the bone in one specimen, and being limited to the anterior half in the other.

The *third* vertebra has a condyle in front and a glenoid cavity behind, and scarcely any indication of a median crest of its neural arch.

The *fourth* vertebra is biconvex, with the centrum much compressed, and passing into a sharp hæmal crest which runs along the whole of its length. There is no neural crest.

The *fifth* vertebra has a glenoid cavity in front and a condyle behind; a median neural ridge is slightly indicated, and, in very old individuals, accompanied by a pair of other ridges diverging in the direction of the posterior zygapophyses. The centrum is compressed into a sharp ridge, highest on the posterior half of its hæmal surface. There is a deep impression inwards and backwards of each of the anterior zygapophyses. The first of the vertebræ, of which the measurements are given below, belonged to an unusually large individual (Plate XI. figs. B).

The *sixth* cervical vertebra (Plate XII. fig. B) has a glenoid cavity in front and a broad condyle behind, which, by a deep median notch, is divided into two lateral halves. Its dorsal surface is flat, without crest, whilst on its visceral surface a rather high crest is evenly continued along nearly the whole length of the vertebra. As mentioned in the preliminary notice, Ann. & Mag. N. H. 1873, vol. xi. p. 307, this vertebra is distinguished by the perforation of the neural arch by a large ovate foramen (*a*) on each side, close to the anterior apophyses. There is no doubt that these two foramina are closed by membrane in fresh specimens, and are the result of the absorption of the osseous substance caused by the pressure of the posterior zygapophysis of the preceding vertebra when the neck is so much bent that the fifth and sixth vertebræ stand nearly at a right angle. I am confirmed in this view by the fact that the natural bend of the cervical column is strongest at this place, which is opposite to the front margin of the carapace. However, although a very deep impression is constantly found immediately behind the anterior zygapophyses, which are perpendicular to the longitudinal axis of the centrum, the actual perforation of the neural arch has not always taken place. It is absent in all the vertebræ of less than 55 millims. length; and out of twelve vertebræ above that size I found it to occur five times.

The *seventh*, biconcave vertebra (Plate XIII. fig. B), is distinguished from that of the Aldabra and Galapagos Tortoises by the entire absence of the high prominence on the dorsal surface of the neural arch. On the other hand, the median crest at the opposite

ɪ

surface of the centrum is as much developed as, or even more than, in any of the species named. A lateral impression behind the anterior zygapophyses is only slightly indicated. Measurements of three specimens will be given below, to show the extent of variation in this vertebra. Specimen *c* belonged to a very large individual, perhaps to the same of which a fifth vertebra has been preserved; and from a comparison with corresponding vertebræ of other gigantic Tortoises we may assume that this individual had a shell about 4½ feet long.

As in the seventh, so in the *eighth* vertebra (Plate XIV. fig. B), the neural crest is either entirely absent or reduced to a pair of low ridges starting from the centre and diverging towards the posterior zygapophyses. In old individuals it is much less distinct than in young ones. The diversity between the stout and slender forms is well marked, as may be seen from a comparison of the subjoined measurements.

	3rd vert.		4th vert.		5th vert.				6th vert.				7th vert.			8th vert.		
	Slen-der.	Stout.	Slen-der.	Stout.	Slender.			Stout.	Slender.			Stout.	*a.*	*b.*	*c.*	Slender. Stout.		
	mm.	mm.	mm.	mm.	mm.	mm.	mm.	mm.	mm.	mm.	mm.	mm.	mm.	mm.	mm.	mm.	mm.	mm.
Length of centrum ..	53	46	85	52	115	80	70	56	110	69	66	71	100	63	58	67	36	30
Depth of centrum in the middle	18	18	20	14	34	21	17	18	27?	23	18	25	50	25	26	44	20	30
Horizontal width of middle of centrum..	11	13	11	8½	22	13	10	12	25	18	17	22	30	19	15	29	16	18
Width of anterior condyle	13	11	13	10	..											..		
Width of anterior glenoid cavity					40	28	25	23	36	26	23	28	55	31	31			
Width of posterior condyle..........			18	13	(broken) 20		18	17	(broken) 26		23	27	..			29	17	16
Width of posterior glenoid cavity	16	16											65	34	36			
Distance of outer margins of anterior zygapophyses	28	27	33	23	45	35	29	30	42	31	30	33	62	40	29	57	28	34
Distance of outer margins of posterior zygapophyses	26	24	27	22	44	29	26	30	(broken) 25		24	27	70	36	33	72	30	40

Of the *dorsal* and *caudal* divisions of the vertebral column only a small number of disjointed vertebræ have been collected. They are of too general a type to require any further remark; and it is quite impossible to form an idea as to the number of the caudal vertebræ.

Limb bones.—In the *scapulary* (Plate XXIV. fig. C) we notice the acute angle at which the scapula and acromium meet; the body of the scapula proper is slender, tri-hedral in form, with its anterior side flat or but little convex. The coracoid becomes ankylosed to the scapula at an early age of the animal; and the neck above its glenoid

surface is very much constricted.   It must be mentioned that there are in the collection
a right and a left scapula, evidently belonging to the same individual, which differ remarkably from all the others in not having the coracoid ankylosed, although the individual appears to have been fully adult, and in having the body of the scapula proper
uniformly compressed.   It is difficult to account for this apparently individual aberration.

The measurements of this bone are the following, a and b being more stoutly built
than c and d :—

	a. millim.	b. millim.	c. millim.	d. millim.
Length of scapula measured from the suture with the coracoid...	205	146	168	135
Circumference in its middle	68	43	42	38
Longitudinal diameter of glenoid cavity	50	35	38	32
Length of coracoid	73	60	66	51
Greatest width of coracoid	70	(?)	54	45
Length of acromium	73	55	65	51

On comparing with each other a considerable number of *humeri* (Plate XXV. fig. C),
they can be arranged in two groups:—1, a slender form, with distinctly trihedral
shaft; to this belong specimens some of which are nine inches in length; other,
smaller ones clearly show by the porous structure of their epiphyses that they belonged
to young or half-grown individuals. 2. The second form of humerus has a shorter and
stouter shaft, and is rather flattened from the front backwards than trihedral. The
specimens of this form do not exceed five inches in length; yet they are clearly bones of
old individuals in which growth had ceased.   As we know that in all the other gigantic
Tortoises, with the life-history of which we are acquainted, the male greatly exceeds in
size the female, it appears to be almost certain that the large bones belonged to the
male and the smaller to the female sex.   In other respects the various humeri exhibit
the following characteristics.   The head is nearly entirely raised above the level of the
summit of the radial tuberosity; the ulnar tuberosity is raised high above the level of
the head, from which it is separated by a deep groove; the ulnar margin of the shaft is
much curved, the opposite margin being nearly straight.   The canal for the blood-
vessels, on the radial edge of the bone close to the elbow-joint, is perfectly bridged over,
perforating the substance of the bone from the front to the hinder side.   The following
measurements are taken from an individual supposed to have been a full-grown male
(a), half-grown male (b), and adult female (c):—

	a. millim.	b. millim.	c. millim.
Length of the humerus (measured in a straight line from the summit of the head to the middle of the trochlea)	210	120	108
Circumference of the narrowest part of the shaft	75	41	44
Longest diameter of the head	39	22	21
Shortest diameter of the head	38	19	19
Extreme breadth between the condyles	64	39	37

The bones of the *forearm* show similar differences as regards their proportions of length and width as the humerus, differences which probably are likewise due to sex. The *ulna* (Plate XXVIII. figs. F & G) is a comparatively straight bone, with its radial edge but slightly emarginate; the *radius* (Plate XXVIII. fig. E) is quite straight and remarkably slender. The ulna of an apparently not quite full-grown male individual (fig. F') has a length of 118 millims., and a width of 14 millims. in its narrowest part; that of an adult female (fig. G) shows only a length of 64 millims. and a width of 11 millims. The longest *radius* in the collection is 105 millims. long, with a circumference of 29 millims. Of the *carpus* and remaining leg-bones nothing has been preserved.

Pelvis (Plate XXVI. fig. C, and Plate XXVIII. fig. A)—In the pelvis the sutures disappear at an early age; the most important point of its structure is the great width of the symphysial bridge between the obturator foramina, which are comparatively narrow. In the majority of the individuals the transverse diameter of this bridge is much greater than the vertical, although the latter is increased by the development of a lower median crest. In very young individuals (that is, in individuals in which the horizontal diameter of the pelvis does not exceed a length of three inches) this characteristic feature of all flat-headed Tortoises is hidden by the development not only of a lower but also of an upper crest. Although the iliac bones are slender, like the remainder of the skeleton, yet the longitudinal diameter of the pelvis does not much exceed the horizontal one. In other respects, only the ilium shows a decidedly characteristic form: from its middle, which is considerably constricted, three ridges arise, running up towards the extremity of the bone, which thus assumes a conspicuously trihedral shape, with three nearly equidistant spinous processes.

The two pelves, of which the following measurements are taken, are nearly perfect; fragments of much larger ones are in the collection.

	millim.	millim.
Longest inner vertical diameter of pelvis from summit of ilium to symphysis	106	80
Longest inner horizontal diameter of pelvis	98	84
Shortest inner horizontal diameter of pelvis (between ilio-pubic prominences)	68	55
Longest diameter of foramen obturatoria	26	21
Width of symphysial bridge	22	18
Depth of symphysial bridge	22	13
Least breadth of posterior portion of ossa ischii	49	38
Length of os ilii	93	70
Least breadth of os ilii	21	15

Femur (Plate XXIX. fig. C).—Although a considerable number of specimens of this bone have been collected, they are, singularly enough, all of middle or small size, and show scarcely any diversity of structure. The shaft is slender, nearly straight, irregularly subtetrahedral, and about as broad in front as behind. The head has an oval form, and is slightly raised above the level of the summit of the larger trochanter, from which it is separated by a deep groove. The larger and lesser trochanters are separated from each other by a more or less deep smooth indentation. I add the measurements of three specimens:—

	a.	b.	c.
	millim.	millim.	millim.
Length	125	104	95
Least circumference..................	45	44	39

Of the *lower leg* no part deserves particular description. To the kindness of M. Bouton I am indebted for a specimen of the right *tibia* of extraordinary dimensions (Plate XXVII. fig. D). It is 200 millims. long, and has in the middle a circumference of 76 millims. The largest specimen in the "Transit-of-Venus" collection is 170 millims. long, with a circumference of 57 millims.

Our two largest specimens of *fibula* measure 140 millims. with a circumference of 41 millims., and 115 millims. with a circumference of 30 millims.

Such of the remains of the *foot* as have been found do not show any peculiarity deserving of notice.

THE RACES OF THE GALAPAGOS.

TORTOISES indigenous to the Galapagos archipelago may be recognized by the following characters :—The nuchal plate is constantly absent; the posterior margins of the two gular plates are convergent, meeting at a more or less obtuse angle, never forming a straight, or nearly straight, transverse line. Neck and legs long. The shell black. One of the scutes on the inner side of the elbow is conspicuous for its size, much larger than those surrounding it.

In the skull, the crown is flat; the palate moderately concave; the front part of the intermaxillary truncated, elevated. The fourth cervical vertebra biconvex. The symphysial bridge between the foramina obturatoria of the pelvis is flat, broader than deep. Osseous carapace very thin. Nuchal vertebræ and limb-bones elongate.

It is evident from this diagnosis that the Galapagos Tortoises are differentiated from the Aldabra races by the same important structural characters as those of the Mascarenes, to which, on the other hand, they are most closely allied in every respect, except in the presence of a double gular plate. The various modifications of the form of the carapace and of the sculpture of the scutes which we have noticed in the forms of the islands of the Indian Ocean we find repeated in those of the opposite end of the globe. PORTER's and DARWIN's statements that the various islands are inhabited by distinct species are fully borne out by my observations. Unfortunately, with the exception of the species collected by Commander COOKSON, we do not possess positive and exact information as regards the localities whence our examples were obtained; but PORTER's accounts are sufficiently detailed to enable us to relegate, with more or less certainty, some of the species before us to the places of their nativity. James Island yielded Tortoises of the broad, circular type; and therefore either *T. elephantopus* or *T. nigrita* came from that island, probably the former. There can be no doubt that we have, in *T. ephippium*, the species inhabiting Charles Island; *T. microphyes* and *T. vicina* we know now to be indigenous to Albemarle, and *T. abingdonii* to Abingdon Island. Possibly other specific forms still exist; but those of Chatham and Indefatigable (and probably Charles) Islands have been extinct for a long time, and not even remains of them are known to exist in collections. As has been said in the " Introduction," the fate of all these animals seems to be sealed: what has been accomplished in the Mascarenes has commenced in the Galapagos.

The specific characters are apparent in mature individuals only. In young examples, which are rather common in collections, the distinctive characters, external or osteological, are incompletely developed, so that it is, at present, extremely difficult and somewhat hazardous to refer very young individuals (up to about 15 inches in length) to the species to which they belong. This resemblance of young examples cannot be used as an argument against the distinctness of the various species, as generally, in Vertebrates as well as Invertebrates, specific characters are not developed before a certain period, which varies exceedingly even in groups nearly related to one another.

1. TESTUDO ELEPHANTOPUS.

The Tortoise to which HARLAN (Journ. Ac. Nat. Sc. Philad. v. 1825, p. 284) gave this name was only 21 inches long over the curvature, or about 17 inches in a straight line, and therefore a young animal. A reference to the measurements and figure given by HARLAN shows clearly that he had an animal with the broad form of the body and with a posteriorly truncated sternum—characteristics by which a small series of examples before me are distinguished, and more especially one individual of nearly the same size as that described by HARLAN.

DUMÉRIL and BIBRON (Erpétol. Génér. ii. p. 115) identify HARLAN's example with one deposited by QUOY and GAIMARD in the Paris Museum under the name of *Testudo nigra*. This specimen is still smaller than HARLAN's, and of an age at which the specific characters are not yet developed; and therefore there is no evidence whatever to show that this identification by DUMÉRIL and BIBRON is correct; and as long as it is uncertain to which of the specific forms the young "*T. nigra*" should be referred, the name had better be disused altogether. DUMÉRIL and BIBRON associate with this young specimen another of large size, distinguished by its broad form, smooth plates, and posteriorly excised sternum, but without giving any convincing proof that these two examples are of the same species. I have not seen an example agreeing in all points with that large example; and it may possibly be another species distinct from those described here.

The materials which I refer to *T. elephantopus* are the following:—

1. An adult male example: a perfect skeleton with carapace, but without epidermoid plates. The carapace is 31 inches long. History of the specimen unknown; purchased of a dealer in Paris for the Oxford Museum, and kindly lent to me by Professor ROLLESTON, F.R.S.

2. An immature female example: a perfect skeleton with carapace, but without epidermoid plates. The carapace is 28¼ inches long. *Hab.* Galapagos Islands. Property of the Royal College of Surgeons. Notes on this example by Professor OWEN in Descript. Catal. Osteol. Ser. R. Coll. Surg. i. 1853, p. 194. no. 1011.

3. Carapace, without epidermoid plates, of an immature male example, 23 inches long. History unknown. Property of the Free Public Museum, Liverpool.

4. Carapace, with epidermoid plates, of a young example, 18 inches long. Sex and history unknown. Property of the Free Public Museum, Liverpool.

5. A perfect skeleton with carapace, 15½ inches long, obtained at Colon, and presented by Captain E. M. Leeds (S.S. 'Tasmanian') to the author, and now in the British Museum.

Carapace.—In the largest example (specimen No. 1) (Plate XXX. fig. A), which has been prepared into a skeleton, the outlines of the epidermoid plates can be clearly traced. It is a fully adult male which, to judge from the condition of the bones, had ceased to grow a long time before its death; the dorsal portion of the shell is extremely thin, in some parts quite transparent. There is almost a total absence of anterior declivity of the first dorsal scute, its front margin being but very little below the level of the highest point of the carapace. The sides of this fore part of the carapace are expanded, not contracted as in *T. ephippium*. The *sternum* is 24½ inches long, and 23 inches broad between the lateral margins of the abdominal plates. It is deeply concave; and when the animal rested on the ground it touched it with the sides of the sternum, which are thicker than the remainder of the carapace, and on a transverse terminal callosity produced by the reverted posterior margin of the sternum, which is straight, truncated, without excision.

Another male example (specimen No. 3) agrees in every respect with the preceding, except in the sexual characters being much less developed, the specimen being only 23 inches long, and therefore much younger. The first dorsal scute is more declivous towards the front, the concavity of the sternum less deep, and its terminal callosity only indicated by the very porous and rough surface of the bone.

In young examples (15 to 18 inches long) the concentric striæ are numerous, but not deeply cut; and in this respect the present species is intermediate between *T. nigrita* and *T. ephippium*. The posterior end of the sternum is nearly truncate, the hind margin of each anal plate being obtusely rounded, and the plates being separated by so shallow a notch that, evidently, with advancing age the sternum would have assumed the same truncate shape which we find in the adult specimens.

It remains to add the principal measurements of the specimens examined:—

	Length of carapace.		Width of carapace.		Sternum.		Caudal plate.	
	In str. line.	Over curve.	In str. line.	Over curve.	Length.	Width.	Length.	Width.
Spec. no.	in.	in.	in.	in.	in.	in.	in.	in.
1. ♂	31	37½	26	40	24½	23		
2. ♀	28½	36½	23	35	22½	19		
3. ♂	23	27½	18	29	18½	16½	1½	3⅜
4. ♀ ? ...	18	22	12½	19¾	14	12	1⅞	3
5. ♀ ? ...	15½	18⅛	11½	19	12½	11	1½	2⅜

Osteology.—The *skull* of an adult example of *Testudo elephantopus* (specimen No. 1, fig. A of Plates XLII.–XLIV.) is distinguished by a very short snout and a singularly raised occipital crest; it is 4⅜ inches long, measured from the front margin of the intermaxillary to the occipital condyle, and 4 inches broad in its widest part, between the zygomatic arches. 1. The frontal region is perfectly flat, broad, passing into the very short snout, its greatest width (in front of the postfrontals) being as much as one half of the distance between the tympanic condyles. 2. The occipital crest is enormously developed; it rises abruptly above the level of the skull, is strongly compressed and scarcely attenuated behind, its extremity being broad and rounded. 3. The tympanic case, with the mastoid, is produced backwards, the hind margin of the paroccipital forming a rather strong curve (fig. A, a). 4. A deep hollow on the lower surface of the occipital in front of the condyle (Plate XLIV. fig. A, b). 5. On the front margin of the temporal fossa, corresponding to the suture between parietal and tympanic, and immediately in front of the foramen carotidis externæ, there is a large, prominent, concave rough tuberosity for the insertion of a portion of the temporal muscle (Plate XLII. fig. A, c); a broad and deep groove (d) separates this tuberosity from the zygomatic arch. 6. Tympanic cavity large, but constricted by the groove just described; the outer tympanic rim is subcircular, with a broad and deep notch (e) in the posterior part of its circumference for the passage of the Eustachian tube. 7. The columella is attached to, and rests upon, a long, straight, sharp ridge, which runs from the notch mentioned to the stapedial foramen. 8. The front margin of the intermaxillary projects beyond that of the frontal, but much less than in T. elephantina; so that the nasal opening, although still obliquely sloping downwards, is as high as broad. 9. The position of the choanæ is advanced forwards; yet, on the palatal view of the skull, a portion of them may be seen uncovered by the alveolar lamellæ of the maxillaries (Plate XXXIX. fig. A). 10. The intermaxillaries are short, one half of the length of the maxillaries; their foremost portion is deeply hollowed out below, and vertically bent downwards to form the truncated beak. The suture between the intermaxillary and vomer is immediately behind the inner angle of the alveolar edges of the maxillaries. 11. The palatal region is much less concave than in the Aldabra Tortoise, and divided along its middle by a high longitudinal crest. The triangular space of which the foramina palatina and the anterior extremity of the vomer form the points is nearly isosceles in shape, in accordance with the generally short longitudinal axis of the skull. Outer pterygoid edge (f) rather elevated and sharp. 12. Anterior surface of the tympanic pedicle deeply excavated.

13. Lower jaw with a double alveolar ridge; its symphysial portion simply vertical, without a backward expansion of the lower margin of the bone. The parts of the angular and coronoid which face each other are closely approximate, leaving only a narrow cleft between them. Upper margin of the angular deeply excised.

The cervical portion of the vertebral column is characterized by its relatively great length. All observers were struck by the length of the neck, which the animal is in the habit of erecting so that the head is raised above the level of the shell. The living animal can turn its head in this position to the right or left, reminding one of a Cobra

K

rising in a posture of defence. This slenderness of the neck is not due to an increase
in the number of vertebræ (which is constant in Tortoises as in Mammals, and limited
to eight), but to their elongated shape. In *T. elephantopus* they are not quite so
slender as in *T. vosmæri*, but much more so than in the species from Aldabra. Also
the spinal canal is narrower than in this latter round-headed form. The crests of the
dorsal as well as those of the hæmal surface are well developed, and sometimes accompanied
by low additional crests. All the articular processes diverge comparatively but little ;
and those which in the Aldabra species are nearly perpendicular to the longitudinal
axis of the vertebræ, are oblique and much depressed in *T. elephantopus*.

In the *atlas* (Plate XLVI. fig. A) the lateral portion of the neural arch (column) is
very much constricted, not broader than the zygapophysis, which is elongate and considerably longer than that part of the bone which forms the roof of the arch. The
centrum (odontoid process) (*a*) is a rhombohedral body.

In the *second* vertebra the neural arch is remarkably compressed and elevated, also
provided with a high neural crest. The *third* has a condyle in front, and a glenoid
cavity behind. The *fourth* is biconvex. The *fifth* (Plate XLVI. fig. C) has a glenoid
cavity in front and a condyle behind ; its median neural crest is low, and accompanied
on each side by two other crests, which diverge in the direction of the posterior zygapophyses. The *sixth* (Plate XLVI. fig. D) has a glenoid cavity in front and a condyle
behind ; its dorsal surface is flat, without crest, whilst on its visceral surface a low crest
is evenly continued along nearly the whole length of the vertebra. The *seventh* (biconcave)
vertebra (Plate XLVII. fig. B) is distinguished by the high crest on its dorsal and visceral surfaces ; in the middle of the vertebra the neural crest is split into two branches,
diverging in the direction of the zygapophyses and leaving a deep triangular recess
between them. The point of divergence forms a kind of summit (*a*) to this vertebra.
The neural arch is deeply hollowed out (*b*) inwards of and behind each anterior zygapophysis to receive the zygapophysis of the preceding vertebra ; but no perforation of
the bone takes place as in the extinct species of Rodriguez. The *eighth* vertebra, with
its bipartite anterior and single posterior condyle, and with its expanded hamate posterior zygapophysis, does not differ from that of the Aldabra species.

The measurements of the second to seventh cervical vertebræ are as follows :—

	2nd. millim.	3rd. millim.	4th. millim.	5th. millim.	6th. millim.	7th. millim.
Length of centrum	55	67	85	83	85	74
Depth of centrum in the middle	34	28	27	27	28	53
Horizontal width of middle of centrum	15	17	18	20	29	27
Width of anterior condyle	15	20	19	...	...	...
Width of anterior glenoid cavity	...	...	...	30	34	40
Width of posterior condyle	...	...	27	32	37	...
Width of posterior glenoid cavity	19	20	...	...	...	43
Distance of outer margins of anterior zygapophyses	23	34	35	38	40	38
Distance of outer margins of posterior zygapophyses	25	26	28	30	29	46

Of the *dorsal vertebræ* scarcely more than the measurements need to be noticed; these are of some importance in comparison with the corresponding vertebræ in other species and also with the cervical vertebræ. The two heads of the first rib are slender, much divergent, leaving a wide triangular space between them and the first dorsal vertebra. The iliac bones abut against the pleurapophyses of the 9th, 10th, 11th, and 12th vertebræ, counting from the first dorsal vertebra. Their distal extremities unite to form the protuberance for the articulation of the ilium.

Length of centrum of dorsal vertebræ:—

1st.	2nd.	3rd.	4th.	5th.	6th.	7th.	8th.	9th.	10th.	11th.	12th.
mm.	mm.	mm.	mm.	mm.	mm.	mm.	mm.	mm.	mm.	mm.	mm.
65	80	80	80	78	55	48	48	16	14	16	22

The number of *caudal vertebræ* is twenty-three in two, and twenty-five in a third; but in the latter specimen they are irregular, and asymmetrically confluent towards the end of the tail.

Limb-bones.—In the *scapulary* (Plate LIII. figs. C, C') we notice the very obtuse angle at which the scapula and acromium meet. The body of the scapula proper is rather slender, compressed, trihedral in form, with its anterior side convex, as shown in the annexed figure, which represents a transverse section through its middle. The coracoid is not ankylosed to the scapula. The measurements of this bone are the following:—

	millim.
Length of scapula (measured from the suture with the coracoid)	200
Circumference in its middle	75
Longitudinal diameter of glenoid cavity	50
Length of coracoid	86
Greatest width of coracoid	70
Length of acromium	84

The shaft of the *humerus* (Plate LI. figs. A, A') is moderately slender, subtrihedral, with the edges well rounded off. There exists a deep impression on the outer side of the bone, immediately below the head and ulnar tuberosity (*a*), and another transverse impression on the hinder side above the trochlea. The ulnar tuberosity projects high above the head, which is nearly entirely raised above the level of the summit of the radial tuberosity. The canal (*b*) for the blood-vessels on the radial edge of the bone, close to the elbow-joint, is perfectly closed, perforating the substance of the bone from the front to the hinder side.

	millim.
Length of the humerus, measured in a straight line from the summit of the head to the middle of the trochlea	216
Circumference of the narrowest part of the shaft	80
Longest diameter of the head	40
Shortest diameter of the head	37
Extreme breadth between the condyles	82

K 2

The bones of the *forearm* do not show any noteworthy peculiarity; but, for the sake of comparison with the other species, I give the measurements:—The ulna has a length of 137 millims., and a width of 28 millims. in its narrowest part; the radius a length of 121 millims., and a circumference of 50 millims., also measured in its narrowest part.

As in *Testudo* generally, so here the *carpal* bones (Plate LIII. fig. D) are arranged in three series, of which the proximal consists of two bones, *lunare* and *cuneiforme*, both articulating with the end of the ulna (*u*); the middle, of the transversely elongate *scaphoid* and "*intermedium*;" and the distal, of five small rounded bones corresponding to, and articulating with, the five metacarpals. The scaphoid articulates with the end of the radius (*r*), the "intermedium" being intercalated between the lunare and third digit. However, in the old specimen of this species there exists the peculiarity that the scaphoid and intermedium are coalesced into a single very long bone (*a*), and that the two radial ossicles of the distal series are similarly united (*b*).

Pelvis (Plate LII.).—In the first place must be noticed the considerable horizontal width of the symphysial bridge (*a*) between the obturator foramina, by which the flat-headed Tortoises are so signally distinguished from the round-headed ones. But quite peculiar to this species is, first, that also the vertical diameter of this bridge is considerable and scarcely less than the horizontal, and, secondly, that, although all other sutures in this aged specimen have disappeared, the transverse suture between the pubic and ischiadic halves of the bridge is still persistent. The iliac bones are comparatively slender, the longitudinal diameter of the pelvis much exceeding the horizontal one. The lower part (*b*) of the pubic bones is gently inclined downwards and slightly concave above; it emits laterally a very long, strong, nearly styliform process (*c*), which is obliquely directed outwards. The posterior part (*d*) of the ossa ischii is of considerable width, very slightly concave above, and provided with a trenchant symphysial crest below, which, expanding hindward, forms a large triangular tuberosity. Lateral margin of the ossa ischii excised in the shape of a C. Obturator foramina of moderate width, considerably wider than the bridge between them, which is not provided above with a median longitudinal crest.

	millim.
Longest inner vertical diameter of pelvis (from summit of ilium to symphysis)	170
Longest inner horizontal diameter of pelvis	132
Shortest inner horizontal diameter of pelvis (between ilio-pubic prominences)	112
Longest diameter of foramen obturatorium	42
Width of symphysial bridge	26
Depth of symphysial bridge	26
Least breadth of posterior portion of ossa ischii	61
Length of os ilii	140
Least breadth of os ilii	30

The shaft of the *femur* (Plate LIII. figs. A, A′, A″) is rather stout, nearly straight, irregularly subtetrahedral, narrower in front than behind. The head has an elliptical form, and does not rise above the level of the summit of the larger trochanter, from which it is separated by a deep and broad cavity. The larger (a) and lesser (b) trochanters are confluent into one broad ridge, and not separated from each other by a smooth groove as we shall find to be the case in some of the following species. The length of the femur in this example is 169 millims., with a least circumference of 80 millims.; the width of the condyles is 66 millims.

Of the *lower leg* no part needs to be mentioned particularly. The *tibia* is 136 millims. long, and the *fibula* 123 millims.

Also the bones of the *foot* may be passed over, with the exception of one point, viz. that, like some bones of the carpus, the astragalus and calcaneum are entirely coalesced, so that no trace of their former separation remains.

2. TESTUDO NIGRITA.

No doubt can possibly be entertained as regards the correct application of this name to the species which I am about to describe. It had been given by DUMÉRIL and BIBRON (' Erpétol. Génér.' ii. p. 80) to two examples, of which the smaller, very young one, is in the Paris Museum, whilst the larger, but also of young age *, is the property of the Royal College of Surgeons. BIBRON's description is almost entirely drawn up from the latter specimen, which, therefore, must be regarded as the type. However, I suspect that the very young example which DUMÉRIL and BIBRON have associated with this specimen should not be referred to this species, but possibly belongs to one of the Mascarene Tortoises. BIBRON, in his description of its legs, omits all mention of the large scute in front of the elbow—a character which, as far as we know at present, is common to all Galapagos Tortoises, but is absent in the Aldabra species. Further, I am almost certain that the large skull described by Dr. GRAY (Shield Rept. p. 6, pl. 34) under the name of *Testudo planiceps* belongs to the present species, for the following reasons:—1. There is the circumstantial evidence that we are acquainted with the adult skulls of *T. elephantopus, T. ephippium, T. abingdonii,* and *T. microphyes,* but not with that of *T. nigrita.* The skulls of the four former species have been preserved, together with their carapaces; but the skull belonging to the shell of our single adult individual of *T. nigrita* is lost. As the skull named *T. planiceps* differs in a marked manner from all the others, we may reasonably suppose that it is that of the last-named species. 2. The British Museum possesses skeletons of young *T. nigrita;* and although the skulls of these individuals have not the specific characters well developed on account of their young age, they show a greater resemblance, especially in its narrower snout, to the skull named *T. planiceps* than to any of the others.

* BIBRON considered it to be an adult example; and its relation to the Galapagos Tortoises appears to have escaped his notice entirely.

The materials available for the description of this species are the following :—

1. A carapace without sternum of a very large example, 41 inches in a straight line ; it was purchased by the Trustees of the British Museum from the Manager of the former Surrey Zoological Gardens, who could not give any information as regards its history (Plate XXX. fig. B).

2. A carapace 22 inches long [*]; type of *Testudo nigrita* (D. & B.); property of the Royal College of Surgeons ; history and sex unknown. I am indebted to Prof. FLOWER, F.R.S., for the loan of this specimen (Plate XXXI. fig. C).

3. Two perfect skeletons, with epidermoid plates, of young examples, the carapaces being 15¼ and 16 inches long. History and sex unknown.

4. A very young example, stuffed ; carapace 8¼ inches long. This specimen was purchased of a collector coming from Chile, and therefore without doubt came originally from the Galapagos Islands. A figure of it, somewhat reduced in size, has been given by Dr. GRAY, under the name of *T. elephantopus*, in Proc. Zool. Soc. 1870, p. 708, pl. 41[†].

5. A very young example, stuffed ; carapace 10 inches long. History and sex unknown. Received from the Haslar Collection.

6. A skull of a very large example, described and figured by Dr. GRAY as *T. planiceps (l. c.)*.

The *carapace* of this species is well characterized by its broad, circular shape, great depth, and more especially by the numerous, deeply cut concentric striæ, by which the areolæ are much reduced in size in immature examples, and which are persistent in considerable number even in specimens of the largest size. Our largest example (specimen No. 1, Plate XXX. fig. B) is a carapace 41 inches long, unfortunately without the sternum. Nevertheless we can safely affirm that this individual was a male, as the females never attain to so large a size. It is only 8 inches longer than broad ; and when measured over the curvature its transverse circumference even exceeds the longitudinal. The areolar portions of the dorsal and marginal plates are perfectly smooth and raised above the general outline of the shell, especially those of the former ; but each plate has a broad margin deeply sculptured with concentric and parallel striæ, the outer striated margin of the marginal plates being even broader than the smooth areolar portion. The first dorsal scute and the anterior half of the second are declivous, the declivity of the former being still steeper than that of the latter.

A deep notch, nearly as deep as that between the two foremost marginal plates, exists between the first and second marginals ; also the posterior margin of the shell is scalloped. The length of the caudal plate is to its width as 11:14 (5½ inches long and 7 inches wide) ; its surface is plane ; that is, its posterior margin is not bent either

[*] BIBRON gives 365 millims. as the length of this example, which is evidently a misprint for 565.

[†] An example of about the same age is rather indifferently figured in SOWERBY and LEAR's 'Tortoises, Turtles, and Terrapins,' where it is named *Testudo indica*.

inwards or outwards. The general colour is a deep black, with a brownish tinge about the margins of the majority of the plates.

As in the preceding species, the shell is thin and light; in this specimen it is only 4 millims. thick in the middle of a costal plate. Specimens of the common *Testudo græca* only about 8 inches long have a carapace almost as thick as these gigantic Tortoises.

The second specimen (Plate XXXI. fig. C), which is 22 inches long and the type of *T. nigrita*, is young, and probably a male, inasmuch as the sternum shows a slight concavity, and the passage between the hind margins of the caudal and sternal plates is of inconsiderable width. As in specimens No. 3 (15½ inches long), the carapace is deeply sculptured all over, the smooth areolæ being very small. Its transverse circumference equals the longitudinal. The front margin, as well as the hind margin, is deeply notched, each notch corresponding to the suture between two marginal plates. The outer surface of the caudal plate is convex, the hind margin being curved inwards; its length is to its width as 3:4. The sternum terminates anteriorly in a thickened, rounded, double-headed transverse knob, with a slightly concave surface below; and posteriorly in a deep rectangular notch. The colour is the same as in the adult example.

Our very young emample, which is only 8¼ inches long, and figured in Proc. Zool. Soc. *l. c.*, agrees in every respect with those of more advanced age, differing from young examples of the same size of *T. ephippium* by the greater relative width of the carapace. The principal measurements of the specimens described are as follows:—

	Length of carapace.		Width of carapace.		Sternum.		Caudal plate.	
	In str. line.	Over curve.	In str. line.	Over curve.	Length.	Width.	Length.	Width.
	in.	in.	in.	in.	in.	in.	in.	in.
1. ♂	41	52	33	53	...	...	5½	7
2. ♂	22	27	16	27	18½	15¾	3	4
3.	15½	19½	11	19	12	9¾	2¾	1[illegible]
	16	21⅓	12½	21	14¼	11½	2⅓	3
4.	8¼	10¼	6	10½	6¾	6	1¼	1

The *skull* (Plates XLII.–XLIV. fig. D) is distinguished by its comparatively longer facial portion, and by the much-produced mastoid processes; it is (see also Gray, Catal. Tort. 1855, 4to, tab. 34) 5¼ inches long, measured from the intermaxillary to the occipital condyle, and 4½ inches broad at its widest part, viz. between the tympanic processes. 1. Its frontal region is flat, narrow, its greatest width being two sevenths of the distance of the tympanic condyles. 2. Only the foremost part of the parietals forms a flat surface, the remainder being compressed into an almost trenchant crest, passing into the long narrow occipital spine, which is scarcely raised above the level of the skull (Plate XLIII. fig. D). 3. The tympanic case with the mastoid is produced far backwards, so that the paroccipital margin appears as a deep semicircular excision

(Plates XLIII. and XLIV. fig. D, *a*). 4. A very deep hollow on the lower surface of
the occipital, in front of the condyle (*b*)*. 5. On the front margin of the temporal
fossa, corresponding to the suture between parietal and tympanic, immediately in front
of the foramen carotidis externæ, there is a large, prominent, flat, rough tuberosity (*c*)
for the insertion of a portion of the temporal muscle; a broad, not very deep groove (*d*)
separates this tuberosity from the zygomatic arch. 6. Tympanic cavity exceedingly
large, especially its posterior portion, the entrance being somewhat narrowed by the
groove just mentioned; the outer tympanic rim is a regular circle, with a shallow notch
in its hinder circumference for the passage of the Eustachian tube. 7. This notch is
very remote from the columellar foramen; and a sharp ridge runs the whole distance
from the notch to the foramen, serving as a rest for the auditory ossicle. 8. The front
margin of the intermaxillaries projects beyond that of the frontals, but much less so
than in the Aldabra Tortoises; so that the nasal opening, although still obliquely sloping
downwards, is scarcely higher than broad. 9. The inner nostrils are advanced, not very
distant from the end of the snout, and on the palatal view of the skull are nearly
entirely hidden below the alveolar lamella of the maxillaries. 10. The intermaxillaries
are short, not quite one half of the length of the maxillaries; and their foremost portion
is deeply hollowed out below, and vertically bent downwards to form the truncated
beak. The suture between the intermaxillary and vomer is immediately behind the
inner angle of the alveolar edges of the maxillaries. 11. Palatal region much less
concave than in *T. elephantina*, and provided with a rather high median longitudinal
crest; posteriorly it is bordered on each side by the raised pterygoid edge, which is
obtuse in its anterior, and trenchant in its posterior half. The distance between the
foramina palatina is much less than their distance from the anterior extremity of the
vomer. 12. Anterior surface of the tympanic pedicle deeply excavated. 13. Lower
jaw with a double alveolar ridge, the symphysial portion being simply vertical, without
a backward dilatation of the lower margin of the bone. The opposite surfaces of the
angular and coronoid are closely approximate, leaving only a narrow cleft between
them. Upper margin of the angular moderately excised.

The skull of a *young* example, $2\frac{1}{3}$ inches long, shows some of the characteristics
described in the adult skull, viz. the greater depth and the less width of the palatal
region, the deep hollow in front of the tympanic pedicle, and the conformation of the
anterior half of the tympanic cavity. The groove between the temporal tubercle and
zygomatic arch, as well as the hollow in front of the occipital condyle, are clearly indicated. On the other hand, the tympanic pedicles are less distant from each other than
in the adult, the mastoido-tympanic process is only slightly produced backwards, and
the occipital crest is short and much less prominent—points of difference which can be
accounted for by the young age of the individual.

* In the figure given by Dr. GRAY the artist has entirely omitted to express the depth of this hollow
by shading.

The description of the skeleton of so young an individual could hardly be accompanied by important results as regards the object of this work, and is therefore omitted.

Caudal vertebræ 24.

3. TESTUDO VICINA.

This race was described by me in the Phil. Trans. 1875, p. 277, from the carapace and skeleton of an adult male example, formerly in the possession of Prof. HUXLEY, and now in the British Museum [*].

The history of that specimen is lost; but, from Commander COOKSON's observations (Proc. Zool. Soc. 1876, p. 524), it seems to be very probable that Albemarle, the largest of the Galapagos Islands, is inhabited by at least two distinct races, and that *T. vicina* is one of them, viz. that of South Albemarle. Having found at the S.W. end, near to Iguana Cove, a Tortoise[†] differing in shape and general appearance from those captured near Tagus Cove (North Albemarle), the shell being deeply striated, and not smooth as is the case in the latter, he remarks :—

" I suppose that abundance or variety of food would be sufficient to produce this difference of appearance in the tortoises from different localities in Albemarle Island, this part of the island, the S.W. end, having a comparatively luxuriant vegetation, whereas the neighbourhood of Tagus Cove is described by Mr. DARWIN as being miserably sterile ; but it is a point not to be overlooked that these two localities are separated by a coast-line of 70 miles, and that between them lie three of the highest volcanic peaks of the whole group, one if not two of which have been active until quite recently ; and from the flanks of all three, streams of black lava descend, each several miles in width. Again, the most northern of these craters is situated almost in the centre of the narrowest part of the island, and sends down its black streams in all directions. . . . I think one may conclude that these lava-streams are quite impassable to the Tortoises, and that, as far as these creatures are concerned, the north and south ends of the island are as effectually separated as they would be by a channel of 40 or 50 miles of deep water."

At the same place where that living specimen was found, Commander COOKSON picked up the skull (without mandible) of what was said to be the largest Tortoise seen

---

[*] My endeavours to trace in the various collections the specimens which are known to have reached England alive within the last forty years have been hitherto singularly unsuccessful; and the present example is the only one which may be supposed to be possibly identical with the individual reported to have been sent to the Zoological Society in 1834, by the Hon. BYRON CARY, from the Galapagos (Proc. Zool. Soc. 1834, p. 113). That specimen is said to have weighed 187 lb., and measured in length, over the curve of the dorsal shell, 44½ inches (I find in our specimen 41½ inches), and along the sternum 25½ inches (as in ours) ; its girth round the middle was 75½ inches (69 inches according to my measurement). It is added that "the lateral compression of the anterior part of the dorsal shell, and the elevation of its front margin . . . are in this specimen strongly marked."

[†] This specimen, unfortunately, is one of those lost in transit to England.

L

for twelve or fifteen years. It is 5½ inches long (to occipital condyle), and agrees in every respect with those from North Albemarle; and as the skulls of all these Albemarle Tortoises are perfectly identical with that of the specimen received from Prof. HUXLEY, which externally differs from the North-Albemarle form by the deep striation of the scutes, we may reasonably infer that this *T. vicina* is actually the race of the southern part of the island.

The form of the carapace (Plate XXXI. fig. A) reminds us of that of *T. elephantopus*; but it is still more depressed, the greater part of the two middle costal plates participating in the formation of the plane surface of the back. The first dorsal scute is but very slightly declivous towards the front; and the edge of the shell along the three anterior marginals is reverted and scalloped: thus the fore part of the shell has in a slight degree the form of a saddle; but it is much less compressed than in *T. ephippium* or *abingdonii*. The striæ of the plates are very distinct, but shallow, and distant from one another (broad), occupying the greater part of the surface of each plate. The striated portions of the plates are not of the same intense black as the smooth ones, but more or less tinged with brown. The shape of the *sternum* differs from that of the preceding species, its gular portion being singularly constricted and having the lateral margins excised. The gular plates are truncated in front. The opposite end of the sternum is dilated, the caudal plates being expanded like wings; their hind margins meet at an obtuse angle. All the plates of the sternum, with the exception of the pectorals and abdominals, are striated, like the dorsal plates. The surface of the sternum is deeply concave.

There is in the British Museum a young stuffed example, with a carapace 12⅛ inches long (without particular indication of its origin), which I am inclined to refer to this species. It has the same depressed shell as the adult, with a similar striation of the plates, and with the anterior margins distinctly reverted; but the sternum is not constricted anteriorly, nor are the caudals expanded like wings. At present we have not the means of judging whether this difference could be accounted for by age or sex; however, as the skull of this young individual agrees singularly well with that of the adult, there is good reason for believing it to be a second example of the same species. The measurements of the two specimens are as follows:—

	Length of carapace.		Width of carapace.		Depth of	Sternum.		Caudal plate.	
	In str.line.	Over curve.	In str.line.	Over curve.	carapace.	Length.	Width.	Length.	Width.
Spec.	in.	in.	in.	in.	in.	in.	in.	in.	in.
Ad. ♂ ...	33	41½	25	42	16	25½	24	4¼	6½
Young...	12¼	14½	9	14½	5½	10½	8½	1¼	2½

The *skull* is so similar to that of *T. microphyes* that it may be noticed in connexion with that species.

Cervical vertebræ.—On comparing the neck-vertebræ of *T. vicina* with those of *T. elephantopus*, we find them generally to be somewhat less slender, and with the crests and ridges less developed; otherwise they are formed according to the same type, and the first, seventh, and eighth are the only vertebræ which exhibit peculiarities indicative of specific distinctness. In the *atlas* (Plate XLVI. fig. B) the lateral portion of the neural arch is but little constricted, at least as wide as the broad zygapophysis, which is longer than that part of the bone which forms the roof of the neural arch. In the *seventh* vertebra (Plate XLVII. fig. C) the summit (*a*) of the neural crest is not single, as in the other species, but split into two prominences, separated from each other by a deep notch. In the *eighth* vertebra the hæmal crest is produced forward to the level of the anterior articular surface, and almost hamate in form, whilst it does not extend beyond the middle third of the length of the centrum in *T. elephantopus*.

Measurements of cervical vertebræ:—

	2nd.	3rd.	4th.	5th.	6th.	7th.
	millim.	millim.	millim.	millim.	millim.	millim.
Length of centrum	47	65	88	80	82	72
Depth of middle of centrum	34	26	25	26	25	49
Horizontal width of middle of centrum	14	18	17	20	25	27
Width of anterior condyle	15	18	20	...	...	...
Width of anterior articular cavity	...	...	...	36	42	41
Width of posterior condyle	...	...	25	28	39	...
Width of posterior articular cavity	19	23	...	...	...	50
Distance of outer margins of anterior zygapophyses	20	33	38	37	42	33
Distance of outer margins of posterior zygapophyses	28	31	31	35	28	55

Dorsal vertebræ.—The last of the three vertebræ which emit pleurapophyses to form the protuberance for the articulation of the ilium is the eleventh; so that only eleven vertebræ can be assigned to this part of the vertebral column. Of the two heads into which the first rib bifurcates, the posterior is more slender than the anterior; the triangular space enclosed by them is wide, but less so than in *T. elephantopus*. For comparison with the latter species I give the length of the centra of the several dorsal vertebræ:—

Dorsal vertebræ	1st.	2nd.	3rd.	4th.	5th.	6th.	7th.	8th.	9th.	10th.	11th.	12th.
	mm.	mm.	mm.	mm.	mm.	mm.	mm.	mm.	mm.	mm.	mm.	mm.
Test. elephantopus	65	80	80	80	78	55	48	48	16	14	16	22
Test. vicina	56	80	87	87	79	61	43	32	17	15	18	(21)

Caudal vertebræ twenty in number; but it is possible that the last rudimentary ossicle has been lost.

Limb-bones.—Singularly enough, the resemblance which we notice between the skulls of this species and *T. ephippium* does not uniformly extend to the other parts of the skeleton, the limb-bones of *T. vicina* being much shorter and stouter than in that species, approaching more *T. elephantopus*. The *scapulary* (Plate LIV. figs. C, C') especially is stout and massive. The angle at which the scapula and acromium meet

L 2

is much less obtuse than in *T. elephantopus* (about 100°); the body of the scapula is compressed, elliptical, with both its anterior and posterior sides equally convex; a transverse section through its middle would be represented by the figure of a greatly elongate O. The shaft of the acromium is trihedral, with the edges rounded, and with the extremity compressed and slightly dilated. The *coracoid* is not ankylosed to the scapula; and its proximal part (neck) is singularly dilated, and very much broader than the corresponding part in *T. elephantopus*. In fact the differences in the scapularies of these two species are so great, that they alone would clearly prove their specific distinctness.

	T. elephantopus, 790 millims. long. millim.	*T. vicina*, 840 millims. long. millim.
Length of scapula (measured from the coracoid suture) ...	200	188
Circumference in the middle of the shaft	75	75
Longitudinal diameter of glenoid cavity	50	55
Length of coracoid	86	83
Greatest width of coracoid	70	74
Least width of neck of coracoid	20	33
Length of acromium	84	78

The *humerus* is so similar to that of *T. elephantopus* (and consequently very dissimilar to that of *T. ephippium*) that no detailed description is needed; but, as in the latter species, the canal for the blood-vessels on the radial edge, close to the elbow-joint, is deep and partly open.

	T. elephantopus. millim.	*T. ephippium*. millim.	*T. vicina*. millim.
Length of humerus	216	285	225
Circumference of the narrowest part of shaft	80	91	95
Longest diameter of the head	40	40	40
Shortest diameter of the head	37	35	38
Extreme breadth between the condyles	82	82	81

The bones of the *forearm* (Plate LIV. fig. D) are also shorter than those of *T. ephippium*, more similar to those of *T. elephantopus*, particularly with regard to the deeply emarginate radial edge of the *ulna*. Both bones are smooth, without prominent ridges or tuberosities. The ulna is twisted round its longitudinal axis, so that the transverse diameters of its proximal and distal dilatations would intersect each other at an angle of about 45°. The olecranon is not much developed. The articular facet of the radius for the articulation of the humerus is a rectangular triangle, with the point directed backwards, and the shortest side in front.

	T. elephantopus, 700 millims. long.	T. ephippium, 840 millims. long.	T. vicina, 840 millims. long.
	millim.	millim.	millim.
Length of ulna	137	155	137
Least width of ulna	28	26	26
Length of radius	121	149	122
Least circumference of radius	50	51	40

Carpus.—The coalescence of the scaphoid and intermedium, and of the two radial ossicles of the third series, which we have found complete in *T. elephantopus*, has commenced in the present individual; but the lines of separation are still clearly visible.

The *pelvis* differs from that of *T. elephantopus* in the same manner as does that of *T. ephippium*; but its horizontal diameter is comparatively greater than in either of those two species. All the sutures are present.

	T. elephantopus, 700 millims. long.	T. ephippium, 840 millims. long.	T. vicina, 840 millims. long.
	millim.	millim.	millim.
Longest inner vertical diameter of pelvis	170	160	157
Longest inner horizontal diameter of pelvis	132	118	144
Shortest inner horizontal diameter of pelvis	112	97	97
Longest diameter of foramen obturatorium	42	42	38
Width of symphysial bridge	26	35	41
Depth of symphysial bridge	26	23	26
Least breadth of posterior portion of ossa ischii	61	80	76
Length of os ilii	140	130	134
Least breadth of os ilii	30	27	29

The *femur* agrees almost entirely with that of *T. elephantopus*, thus differing from that of *T. ephippium* in the same points, which will be indicated in the description of the latter species. The bones of the *lower leg* and *tarsus* do not show any noteworthy peculiarity, the state of coalescence of the astragalus and calcaneum being the same as in some of the carpal bones mentioned above.

	T. elephantopus, 700 millims. long.	T. ephippium, 840 millims. long.	T. vicina, 840 millims. long.
	millim.	millim.	millim.
Length of the femur	169	186	165
Least circumference of the femur	80	90	79
Width of the condyles	66	67	73
Length of the tibia	136	150	129
Least circumference of the tibia	60	72	57
Length of the fibula	123	138	123
Least circumference of the fibula	45	45	43

4. Testudo microphyes.

This race was described by me in Phil. Trans. 1875, p. 275, from a specimen $22\frac{1}{2}$ inches long, then lent to me by the Museum Committee of the Royal Institution of Liverpool, but since acquired by the Trustees of the British Museum.

In that original description I stated that it was a fully adult male, drawing this inference from the concavity of its sternum—and suggested, moreover, that the specimen, on account of its small size, might be the Hood's-Island race of PORTER. The specimens collected by Commander COOKSON in North Albemarle prove at least the first of these views to have been erroneous, the typical specimen being undoubtedly a female. In this species, and perhaps also in the other flat-headed Tortoises, the sternum of the adult female is conspicuously concave, much more so than in males of the same *size*, though not of the same *age*. The concavity of the sternum of females extends over a greater portion of the surface of the bone than in the males (in which it is deepest behind the middle), and never attains the same degree of depth; also the sides of the sternum and the caudal extremity never become swollen and callous.

Commander COOKSON obtained his specimens near to Tagus Cove, about 4 miles inland, on a small elevated plateau covered with stunted bush and high, very coarse grass. Of twenty-four examples of all ages the following reached England [*] :—

A. Adult male, living; carapace $33\frac{1}{2}$ in. long (in a straight line); weight 240 lb.

B. Adult male, shell, skull, and skin of legs; carapace $33\frac{1}{2}$ in. long.

C. Half-grown male, shell, skull, and skin of legs; carapace 25 in. long.

D. Young male, living; carapace $16\frac{1}{2}$ in. long. Brought home on board the 'Challenger.'

E. Adult female, living; carapace 27 in. long.

F. Adult female, living; carapace $25\frac{1}{2}$ in. long. Brought home on board the 'Challenger.'

G. A very young specimen, living; carapace 18 in. long.

A very great advantage has been gained by Commander COOKSON bringing home several individuals from the same locality, thus enabling us to discriminate between specific characters and other modifications due to various causes. And it was important to ascertain that there is no variation whatever either in the three skulls from Albemarle Island or in those from Abingdon, and that, likewise, the carapaces from the same island differed only in those points which I have always considered to be due to sex or subject to change with the progress of growth.

As long as the Tortoises are young, growth, as far as it is externally visible, proceeds along the margins of all the scutes; the sutures get broader, appearing as whitish seams, soft and very sensitive. After some time the young portion of the epidermis

---

[*] The living specimens have been temporarily deposited in the Zoological Gardens; two were put by Commander COOKSON on board the 'Challenger.'

becomes horny, and is raised in a line (stria) running along each side of the suture. At a later period this increment takes place only (or, at least, only conspicuously) in certain portions of the carapace. Thus, in those species in which the male is distinguished from the female by a saddle-like compression of the front portion, the sutures between the first two costals and the corresponding marginals and pectoral widen, the change in the form of the carapace being almost entirely confined to this portion. It is impossible to say at what period growth ceases in these animals; probably it greatly varies in different individuals. The gigantic specimen of *Testudo elephantina*, which I observed in a living state, and which was certainly 100 years old, if not more, was still growing at the time of its death; but the outward signs of the increment of the shell were confined to the upper and lower margins of the lateral marginal scutes; and the animal may have gained thereby an increase of one inch in the circumference of its carapace within the period of a year.

Description of the Male.—In one of the largest male specimens (A, 33¼ inches) (Plates XXXIV. & XXXV. fig. A) the shell is moderately depressed, with the upper surface flattened, the upper profile from the centre of the fourth vertebral scute being nearly straight. However, the central line of the first vertebral, which is raised into a slight ridge, gently slopes downwards towards the front. Each side of the fore part of the carapace is deeply concave; but the margins are but slightly reverted and scarcely scalloped. The part of the shell above the hind legs is bulged outwards, especially towards the margin, all this reminding us of that peculiar form of shell which we find fully developed in the Charles and Abingdon, and in one of the Aldabra Tortoises. The declivity of the last vertebral and caudal is very steep; and the latter scute is slightly bent inwards. The entire surface of the carapace is irregularly pitted from the unequal development of the osseous substance; and nearly every trace of striation has disappeared. The two front marginal scutes join each other by a very short suture. The colour is a dark horn-colour, almost black.

The head, of which I give a full-size figure, in comparison with that of *T. elephantina* (Plates XXXII. & XXXIII.), is comparatively much larger than in the latter species, and can be held nearly at a right angle to the perpendicular neck; the animal is able to turn it aside sufficiently to look at objects behind it. The neck is often held erect, raised high above the carapace, but seems to be not quite so long as in the Abingdon Tortoise.

In the second large male specimen (B, 33⅓ in. long), the shell is somewhat less depressed, the areolar centres of the three middle vertebrals being distinctly raised; and the upper profile slopes downwards from the centre of the second vertebral towards the front. The sides of the fore part of the carapace are less concave; a deep indentation between the first and second marginals. The posterior declivity of the shell is very steep from the very projecting areolar centre of the penultimate vertebral. The surface of this carapace is very smooth, with only a few shallow pits. Sternum deeply concave, with short and broad anterior and posterior lobes, and truncated in front and behind.

In a half-grown male (C, 25 in. long) (Plate XXXVIII.) the reversion of the margins is only slightly indicated. The striation occupies a broad band round the periphery of each scute; but the striæ are very superficial. The concavity of the sternum is very shallow; and there is a very open indentation on the hind margin of the caudal scutes.

Description of the Female (Plate XXXV. fig. A, and Plates XXXVI. & XXXVII.).—The carapace is more depressed than in the male; and the upper profile slopes from the middle towards the front, being quite a steep decline along the first vertebral scute. There is scarcely any concavity on the sides of the fore part of the carapace; and the part of the shell above the hind legs is but slightly bulged outwards. The two front marginal scutes join each other by a much longer suture than in the male. The caudal plate is distinctly turned outwards. Sternum slightly concave, and caudal margin somewhat indented, as in the half-grown male specimen (C). The surface of the carapace is pitted, without any trace of striæ. Tail very short (only 2 inches long).

In the typical specimen (Plate XXXVII.), which, as mentioned above, is likewise a female, the plates are not pitted, and the sternum is truncated behind. As a (probably) individual peculiarity must be noticed the confluence of the two anterior marginals into one plate on each side.

The females erect their neck much less frequently, and never to the same extent as the males; whether the vertebræ themselves are comparatively shorter than in the male cannot be ascertained for the present.

Very young specimens (up to 16 inches in length) show a very well-marked striation of the whole surface of the scutes, the areola alone excepted, and cannot be distinguished from the young of other species.

Measurements.

	Spec. typ., ♀. in.	A, ♂. in.	B, ♂. in.	C, ♂. in.	E, ♀. in.
Length of carapace in straight line	22½	33½	33½	25	27
Length of carapace over curvature	26	44	41	32½	35
Width of carapace in straight line	15½	25	24½	19½	22
Width of carapace over curvature	29	46	42	34½	38
Length of sternum	18	27½	24½	20½	21½
Width of sternum	14	25	23	16½	20
Width of postgular plates	8	14	12	10	10
Length of postgular plates	4½	9	6	5	6
Length of caudal plate	2	4½	3¾	3	4¾
Width of caudal plate	3⅜	8	6½	3¾	2¾

Skull (Plate XLV. figs. A–C, ♂: Plates XLII.–XLIV. fig. B, ♀).—The principal differences in the skulls of the Galapagos Tortoises are found in the palate and lower part of the skull. The Albemarle races, *T. microphyes* as well as *T. vicina*, have the outer pterygoid expanded, differing therein from *T. elephantopus* and *T. nigrita*, and agreeing in this respect with *T. ephippium*, in which, however, the expansion is excessive and the concavity less. Our largest *male* skull of the North-Albemarle race (Plate XLV. figs. A–C) is $4\frac{3}{4}$ inches long: in this skull the foramina palatina are 29 millims. distant from each other, and the posterior roots of the pterygoid edges 25 millims.; so that the convergence of the latter is very gradual. The distance of a foramen palatinum from the anterior extremity of the vomer is 30 millims. The ridges on the base of the skull are obsolete; but the impression in front of the occipital condyle is deep. Parietal crest raised above the surface of the cranium, of moderate length. Articular surface of the mandible very narrow.

In all these particulars our skull of *T. vicina* (Plate XLVII. fig. A) agrees with that of the northern form. It is also that of a male, $4\frac{7}{8}$ inches long, measured from the front margin of the intermaxillary to the occipital condyle, and 4 inches broad in its widest part, between the zygomatic arches. Its outer tympanic rim has a subcircular outline ; and the ridge which runs from the Eustachian notch (*e*) to the stapedial foramen, and to which the columella is attached, is high and rather sharp.

Of the skull of the *female* (Plates XLII.–XLIV. fig. B) I have given a detailed description (Phil. Trans. 1875, p. 276). It differs from that of the male in having the occipital crest and the tympanic case but little produced backwards, being in general configuration but little removed from the young skull, although it is one of a fully adult animal. In the depth and width of the mesopterygoid region, in the expansion and direction of the pterygoid edges, there is no difference whatever from the male.

5. TESTUDO EPHIPPIUM.

I proposed (Phil. Trans. 1875, p. 271) this name for a species equally well characterized by the peculiar form of its carapace and of its skull. PORTER's remarks on the Tortoises of Charles Island (see *ante*, p. 5) apply so well to this species that I have no doubt that the specimen from which the following description is taken came from that island. If this is really the case, this species is probably extinct. The specimen is an adult male, 33 inches long, stuffed, and belongs to the Museum of Science and Arts, Edinburgh. It was lent to me by T. C. ARCHER, Esq., Director of the Museum, who most kindly allowed the skull and limb-bones to be extracted, which could be effected without the least injury to the outward appearance of the specimen. Nothing is known of its history.

A very young stuffed example, 7 inches long, in the British Museum is referred to this species on account of its oblong shape and large smooth areolæ.

The *carapace* (Plates XXXIX. and XXXI. fig. B) is narrow, oblong, and deep ; from the middle of the central dorsal plate to the front margin of the shell the upper profile

M

is nearly horizontal, the fore part of the shell being strongly compressed, concave on each side, with the anterior margin strongly reverted—this part of the shell having an appearance which has been so aptly compared by Porter to a "Spanish saddle." The hind part of the shell is rounded, with a steep posterior profile, but more gently declivous on the sides, the marginal plates above the hind legs being arched outwards with the edge somewhat reverted, but less so than on the anterior marginal plates. The anterior as well as the posterior margins are irregularly scalloped. The plates are nearly smooth, the areolar portions passing gradually into the striated portions; but the striæ themselves are inconspicuous and in many places nearly obliterated. The sternum * is deeply concave, truncated in front and behind, the substance of the caudal plates and of the lateral portion of the abdominals being much thickened.

I need not mention the scutellation of the head and legs, none of the Galapagos Tortoises showing any peculiarity in this respect. The tail is very short, and without terminal "claw."

On comparing the carapace of the young example with that of equally small specimens of other species, we find the areolar spaces larger, the concentric striæ deeply sculptured, but less numerous and further apart. Especially the sternal plates are smooth, with the striæ partly obliterated. Posteriorly the sternum terminates in a notch (and this appears to be uniformly the case in very young specimens of all the species); but this notch is much shallower than in *T. nigrita*, obtuse-angular.

The measurements of these two specimens are the following:—

	Length of carapace.		Width of carapace.		Depth of carapace.	Sternum.		Caudal plate.	
	In str.line.	Over curve.	In str.line.	Over curve.		Length.	Width.	Length.	Width.
Spec.	in.	in.	in.	in.	in.	in.	in.	in.	in.
Adult	33	40	23½	40	17	2¼	21¼	3½	6
Young	7	9½	5	8½	3½	4¾	6¾	1	1¼

Skull.—The skull (Plates XLII.–XLIV. fig. C) is comparatively smaller than that of *T. elephantopus*; it is 4¾ inches long, measured from the front margin of the inter-maxillary to the occipital condyle, and 3¾ inches broad in its widest part between the zygomatic arches. The sutures between the various bones can be clearly traced; and growth evidently had not ceased entirely—an observation confirmed by the examination of other bones extracted from the specimen. 1. The frontal region is flat, broad, passing into the very short snout, its greatest width (in front of the postfrontals) being about one half of the distance between the tympanic condyles. 2. The occipital crest is moderately developed, pointed behind, and rising but little above the level of the upper surface of the skull. 3. The tympanic case with the mastoid is produced back-wards, the hind margin of the paroccipital forming a rather strong curve (Plate XLII.

* A large portion in the middle of the sternum has been cut out by the person who preserved the animal, in order to extract the contents of the shell.

fig. A, *a*). 4. There is no hollow in front of the occipital condyle, the space between the condyle and basisphenoid simply shelving downwards towards the latter (*b*). 5. On the front margin of the temporal fossa, in front of the *foramen carotidis externæ*, there is a large not very prominent tuberosity (*c*) for the insertion of a portion of the temporal muscle; no groove separates this tuberosity from the zygomatic arch; or, in other words, the tympanic cavity is not constricted in front. 6. Tympanic cavity very large: the outer tympanic rim ovate, resembling the outline of the human concha, with the convex side in front and the pointed part above; the notch for the passage of the Eustachian tube is very broad, but shallow (*e*). 7. The ridge which runs from this notch to the stapedial foramen, and to which the columella is attached, is rather low and obtuse. 8–10. The points noticed under these figures in the description of the skull of *T. elephantopus* (see page 65) are exactly the same in the present species. 11. The palatal region is very shallow and broad, in consequence of the outer pterygoid-edge being flattened down and expanded in its whole length (Plate XLIV. fig. C, *f*). The triangular space, of which the foramina palatina and the anterior extremity of the vomer form the points, is isosceles in shape, in accordance with the generally short longitudinal axis of the skull. 12. Anterior surface of the tympanic pedicle with a shallow impression. 13. Lower jaw with a double alveolar ridge; its symphysial portion simply vertical, without a backward expansion of the lower margin of the bone. The parts of the angular and coronoid which face each other leave a rather wide cleft between them. Upper margin of the angular not excised.

Limb-bones.—The following bones have been preserved in the large stuffed example, and were extracted from it:—The *humerus* (Plate LI. figs. B, B'), distinguished by its great length and slenderness; its shaft is trihedral in the middle, and not much bent. The two hollows which we noticed in *T. elephantopus* below the head and above the trochlea are here absent. The ulnar tuberosity (*a*) projects high above the head, which is nearly entirely raised above the level of the summit of the radial tuberosity. The canal (*b*) for the blood-vessels on the radial edge, close to the elbow-joint, is deep and partly open, cutting off, as it were, a splint from the radial extremity of the bone.

	T. elephantopus, 700 millims. long. millim.	*T. ephippium,* 840 millims. long. millim.
Length of the humerus (measured in a straight line from the summit of the head to the middle of trochlea)	216	235
Circumference of the narrowest part of the shaft	80	91
Longest diameter of the head	40	40
Shortest diameter of the head	37	35
Extreme breadth between the condyles	82	82

The bones of the *forearm* (Plate LIV. fig. B) are, like the humerus, comparatively slender; they are remarkably smooth, without prominent ridges or tuberosities. The ulna has its radial edge but slightly emarginate, and is twisted round its longitudinal axis, so that the transverse diameters of its proximal and distal dilatations would intersect each other at an angle of about 50°. The olecranon is not much developed. The articular facet of the radius for the articulation with the humerus is a rectangular triangle, with the point directed backwards, and the shortest side in front.

	T. elephantopus, 700 millims. long. millim.	T. ephippium, 840 millims. long. millim.
Length of ulna	137	155
Least width of ulna	28	26
Length of radius	121	149
Least circumference of radius	50	51

Only a few of the *carpal* bones have been extracted from the specimen, among them the scaphoid and "intermedium," which have remained perfectly separate.

In the *pelvis* (Plate LIV. fig. A) we notice, in the first place, that all the sutures are present, and that growth was still proceeding in their vicinity. However, on the whole, the pelvis does not differ in a marked manner from that of *T. elephantopus*, except that the symphysial bridge is broader (the obturator foramina consequently narrower) and not so deep. The posterior part of the ossa ilii also is broader than in the other species. Other slight differences of form may be seen from the accompanying comparative measurements :—

	T. elephantopus, 700 millims. long. millim.	T. ephippium, 840 millims. long. millim.
Longest inner vertical diameter of pelvis (from summit of ilium to symphysis)	170	160
Longest inner horizontal diameter of pelvis	132	118
Shortest inner horizontal diameter of pelvis (between iliopubic prominences)	112	97
Longest diameter of foramen obturatorium	42	42
Width of symphysial bridge	26	35
Depth of symphysial bridge	26	23
Least breadth of posterior portion of ossa ischii	61	80
Length of os ilii	140	180
Least breadth of os ilii	30	27

The *femur* is very similar to that of *T. elephantopus* (p. 68), with the exception of its proximal portion (Plate LIII. fig. B): the head has an elliptical form, and does not rise above the level of the summit of the larger trochanter, as in *T. elephantopus*, but is considerably smaller; on the other hand, the cavity separating the head from the trochanters is much larger, as broad as long, and the two trochanters (*a* and *b*) are widely separated from each other by a smooth groove.

The bones of the *lower leg* and *carpus* do not show any noteworthy peculiarity. As in *T. elephantopus*, the astragalus and calcaneum are coalesced; but, owing to the less-advanced age, the line of separation is still visible.

	T. elephantopus, 790 millim. long. millim.	*T. ephippium*, 940 millim. long. millim.
Length of the femur	109	186
Least circumference of the femur	80	90
Longest diameter of head of femur	55	43
Width of the condyles	66	67
Length of the tibia	136	150
Least circumference of the tibia	60	72
Length of the fibula	123	138
Least circumference of the fibula	45	45

6. TESTUDO ABINGDONII.

As mentioned above (pp. 5, 6), our knowledge of the existence of these animals in Abingdon was derived from two short statements by Captains JAMES COLNETT [*] (1798) and BASIL HALL [†] (1822). Since the time of these voyagers down to the present date no further information has been received; and no specimen seems ever to have been brought to Europe before Commander COOKSON's visit to the island. He succeeded in bringing home the following specimens [‡], all of which are adult males:—

1. Carapace, 38 inches long, with skull and skin of neck, limbs, &c. Weight 201 lb.
2. Stuffed specimen with skull; carapace 34 inches. Weight 131 lb.
3. Stuffed specimen with skull; carapace 38½ inches.

These three examples are entirely identical with regard to the form of the carapace, as well as to the structure of the skull and the other parts which have been preserved. They resemble externally *T. ephippium*; but the carapace is of extraordinary thinness and lightness, with the surface deeply pitted and grooved from the deficiency of osseous substance, which is quite in accordance with the slenderness of the cervical vertebrae. For this reason these Tortoises must be much more liable to injury from blows, falls, and other rough treatment; and no doubt Commander COOKSON (P. Z. S. 1876, p. 523) is quite right in supposing that it is owing to this cause that he could not keep them alive on board. The skull, however, does not participate in this general atrophied condition of the skeleton, but is as stout and compact as in the other species.

Commander COOKSON has already pointed out some of the most striking distinctive

[*] 'Voyage to the South Atlantic' (London, 1798), p. 152.
[†] 'Extracts from a Journal' (Edinburgh, 1824), 2nd edit. ii. p. 140.
[‡] The particulars of their capture are related by Commander COOKSON in Proc. Zool. Soc. 1876, p. 520.

features of this species, of which the white colouring of the jaws and upper part of the head and of some of the nails is not the least singular; it is, perhaps, an instance of commencing albinism. The circumstance that no young individuals or females were found must be accounted for by the greater difficulty of discovering small individuals, as well as by the well-known fact that the females are much less active, and more disinclined to leave their hiding-places, than the males.

The anterior portion of the carapace (Plates XL. & XLI.) is narrow, compressed, and deep; the posterior more depressed and dilated. The upper profile is almost straight, and ascending forwards almost from the areolar centre of the last scute, the highest point of the shell being quite in front, so that the anterior aperture assumes the shape of a triangle pointed at the apex. The fore part of the shell is strongly compressed, concave on each side, with the anterior margin strongly reverted, having the appearance of a "Spanish saddle" in a still more marked degree than in *T. ephippium.* Posteriorly the carapace is dilated, with rounded posterior margin, with a rather steep decline over the caudal plate, which is bent inwards, but more gently declivous on the sides, the marginal plates above the hind legs being strongly bulged outwards, and the posterior marginals being slightly reverted. The anterior margins of the shell are irregularly scalloped and more or less broken, as it is this part of the shell chiefly with which the males butt each other in their combats. The posterior margin of the carapace shows scarcely any indentation. The plates are without any trace of striation; but the whole surface of the shell is deeply and irregularly pitted.

The sternum is deeply concave, truncated in front and behind, with the front lobe rather contracted.

The jaws, some of the nails, and portions of the sternum are of a yellowish white colour. The tail is not very long, and without terminal claw.

Length of carapace.		Width of carapace.		Depth of	Sternum.		Postgular plates.	
In str. line.	Over curve.	In str. line.	Over curve.	carapace.	Length.	Width.	Length.	Width.
in.	in.	in.	in.	in.	in.	in.	in.	in.
38	40½	22½	38	19½	26	21½	7	12

Skull (Plate XLV. figs. D–F).—The three skulls of this Tortoise are perfectly identical, showing more especially the same singular formation of the palate. The palate is narrow and very concave, bordered by the expanded pterygoid-edges, which rapidly converge behind. In a skull 4⅜ inches long the foramina palatina are 23 millims. distant from each other, and the posterior roots of the pterygoid-edges 16 millims. The distance of a foramen palatinum from the anterior extremity of the vomer is 32 millims. The inner margin of the pterygoid-edge is continued backwards as a sharp ridge over the base of the skull. The impression in front of the occipital condyle is deep. Parietal crest almost in a line with the upper surface of the skull, and long. Articular surface of the mandible much dilated.

Cervical Vertebræ.—Of the specimen of which the carapace measures 38 inches in length, and the skull 4⅜ inches, the cervical vertebræ (with the exception of the first) have been preserved; in their natural connexion they measure 25 inches, which gives an idea of the slenderness of the neck. They exceed in slenderness those of all the other Galapagos Tortoises as far as we know them at present, and as will be seen from the list of measurements subsequently given. They are of extraordinary lightness compared with their size and length, perfectly normal, and not diseased, as Commander Cookson believed on his first inspection of the animal after its death (*l. c.* p. 521).

In the *second* vertebra (Plate XLVIII. figs. A) the neural arch is compressed and elevated, and provided with a neural crest, the anterior summit of which is expanded. The *third* (Plate XLVIII. figs. B) has a condyle in front, and glenoid cavity behind; its neural crest is low, and split into three ridges; hæmal crest well developed. The *fourth* (Plate XLVIII. figs. C) is biconvex, much longer than the third, with low crests. The *fifth* (Plate XLIX. figs. D) has a glenoid cavity in front and a condyle behind, and is bulkier, but scarcely longer than the fourth; on its neural aspect it presents a pair of sharp low ridges, confluent in front and divergent towards the posterior zygapophyses, the median line being provided with another still lower ridge; hæmal crest very low. The *sixth* (Plate XLIX. figs. E) is about as long and bulky as the fifth, and has a glenoid cavity in front and a condyle behind, both the glenoid cavity and the condyle showing indications of a bipartition. The neural arch is compressed, with a pair of very low, parallel, close ridges arising from the roots of the zygapophyses, and disappearing towards the middle of the bone. The hæmal surface has a higher crest than any of the preceding vertebræ. The *seventh* vertebra (Plate L. figs. F) has a glenoid cavity in front and behind, and the bipartition of the posterior is nearly complete; the neural crest is very high, and split into two branches diverging backwards in the direction of the zygapophyses, and leaving a deep triangular recess between them; as in *T. vicina*, the summit of this crest is split into two prominences, separated from each other by a deep notch. Whilst in the four preceding vertebræ the neural arch is deeply hollowed out inwards of and behind each anterior zygapophysis, this place is rather convex in the seventh vertebra of this species. The hæmal crest is as developed as in the preceding vertebra, but higher. The *eighth* vertebra (Plate L. figs. G), with its bipartite anterior and single posterior condyle, and with its expanded hamate posterior zygapophyses, does not essentially differ from that of the other species; but all the grooves and impressions are still more strongly marked.

The following Table will show the measurements of these vertebræ compared with those of *T. elephantopus* and *T. vicina*:—

Vertebræ	2nd.			3rd.			4th.			5th.			6th.			7th.			8th.		
	T. elephantopus.	T. vicina.	T. abingdonii.	T. elephantopus.	T. vicina.	T. abingdonii.	T. elephantopus.	T. vicina.	T. abingdonii.	T. elephantopus.	T. vicina.	T. abingdonii.	T. elephantopus.	T. vicina.	T. abingdonii.	T. elephantopus.	T. vicina.	T. abingdonii.	T. elephantopus.	T. vicina.	T. abingdonii.
	mm.	mm.	mm.	mm.	mm.	mm.	mm.	mm.	mm.	mm.	mm.	mm.	mm.	mm.	mm.	mm.	mm.	mm.	mm.	mm.	mm.
Length of centrum	55	47	62	67	65	88	85	88	122	83	80	114	85	82	118	74	72	96	..	45	63
Depth of centrum in the middle	34	34	34	28	26	34	27	25	33	27	26	34	28	25	38	53	49	68		34	46
Horizontal width of middle of centrum	15	14	15	17	18	18	18	17	23	20	20	22	29	25	29	27	27	26		24	30
Width of anterior condyle	15	15	24	20	18	22	19	20	25							..	..		..	46	45
Width of anterior glenoid cavity	..									30	36	34	34	42	37	40	41	44			
Width of posterior condyle	..						27	25	31	32	28	31	37	39	42	..	..		29	39	
Width of posterior glenoid cavity	19	19	21	20	23	26										43	50	47	..		
Distance of outer margins of anterior zygapophyses	23	20	28	34	33	39	35	38	38	38	37	43	40	42	53	38	33	44	..	54	60
Distance of outer margins of posterior zygapophyses	25	28	35	26	31	31	28	31	37	30	35	50	29	28	36	46	55	56		65	76

EXPLANATION OF THE PLATES.

PLATE I.
Testudo elephantina, adult ♂, nat. size (spec. *c*). Side view of head.

PLATE II.
Testudo elephantina, adult ♂. Upper view of head.

PLATE III.
Testudo elephantina ♂, ⅓ nat. size (spec. *a*).

PLATE IV.
Lower views of carapaces.
Fig. A. *Testudo elephantina* ♂.
Fig. B. *Testudo daudinii* ♂.

PLATE V.
Testudo daudinii ♂, ⅕ nat. size.

PLATE VI.
Testudo ponderosa ♀, ⅕ nat. size.

PLATE VII.
Testudo hololissa ♂, ⅓ nat. size (from a specimen in the Royal College of Surgeons).

PLATES VIII. & IX.
Fig. A. Skull of *Testudo elephantina*, nat. size (spec. *n*).
Fig. B. Skull of *Testudo ponderosa*, nat. size.

PLATE X.
Cervical vertebræ of *Testudo elephantina*, nat. size (spec. *c*).
Fig. A. Two views of atlas.
Fig. B. Three views of second vertebra.
Fig. C. Three views of third vertebra.

N

PLATE XI.

Fifth cervical vertebra (nat. size) of:—

Fig. A. *Testudo elephantina* (spec. *c*).

Fig. B. *Testudo vosmæri*.

PLATE XII.

Sixth cervical vertebra (nat. size) of:—

Fig. A. *Testudo elephantina*.

Fig. B. *Testudo vosmæri*.

PLATE XIII.

Fig. A. Seventh cervical vertebra (nat. size) of *Testudo elephantina*.				.

Fig. B. Seventh cervical vertebra (nat. size) of *Testudo vosmæri*.

Fig. C. Atlas of *Testudo ponderosa*.

Fig. D. Upper (D), lower (D'), and front (D'') views of ninth caudal vertebra of *Testudo elephantina*.

PLATE XIV.

Fig. A. Eighth cervical vertebra of *Testudo elephantina*.

Fig. B. Eighth cervical vertebra of *Testudo vosmæri*.

Fig. C. Upper (C) and lower (C'), front (C'') and back (C''') views of eighteenth caudal vertebra of *Testudo elephantina*.

PLATE XV.

Caudal vertebræ of *Testudo elephantina*.

Fig. A. Upper, lower, and side views of first.

Fig. B. Upper, lower, and side views of fourth.

Fig. C. Upper, lower, and side views of sixth.

PLATE XVI.

Humerus of *Testudo elephantina* (spec. *c*), front and back views.

PLATE XVII.

Pelvis of *Testudo elephantina*, ½ nat. size (spec. *c*).   Front, side, and upper views.

PLATE XVIII.

Pelvis of *Testudo ponderosa*, nat. size.   Front, side, and upper views.

PLATE XIX.

Femur of *Testudo elephantina*, nat. size (spec. *c*), four views.

	a. Larger, *b*. Lesser trochanter.

PLATE XX.

Carapace of *Testudo inepta*, ¼ nat. size.

Fig. A. Side view.   Fig. B. Dorsal view.   Fig. C. Posterior view.

Carapace of *Testudo triserrata*, from a specimen in Mr. Newton's collection.

Fig. D. Side view, restored from three fragments (⅙ nat. size).

Fig. E. Sternum (¼ nat. size).

PLATE XXI.

Testudo vosmæri ♂, ⅔ nat. size, partly restored.

PLATE XXII.

Testudo vosmæri ♀ juv., ⅓ nat. size.

PLATE XXIII.

Skulls (nat. size) of :—

Fig. A. *Testudo triserrata.*

Fig. B. *Testudo inepta* (Newton Collection).

Fig. C. *Testudo vosmæri.*

PLATE XXIV.

Scapula (nat. size) of :—

Fig. A. *Testudo ponderosa.*

Fig. B. *Testudo triserrata.*

Fig. C. *Testudo vosmæri.*

PLATE XXV.

Humerus (nat. size) of :—

Fig. A. *Testudo triserrata* : posterior view of spec. No. 39942.

Fig. B. *Testudo inepta* : posterior view of spec. No. 39943.

Fig. C. *Testudo vosmæri* : posterior and anterior views.

PLATE XXVI.

Pelvis : anterior view (nat. size) of :—

Fig. A. *Testudo triserrata* ♂ : Newton Collection.

Fig. B. *Testudo inepta* ♂ : spec. 75. 10. 20. 1.

Fig. C. *Testudo vosmæri.*

PLATE XXVII.

Fig. A. Upper view of pelvis of *T. inepta* (75. 10. 20. 1).

Fig. B. Upper view of pelvis of *T. vosmæri* (76. 11. 1. 113).

Fig. C. Front and end views of tibia of *T. triserrata* or *inepta* (No. 39961).

Fig. D. Front view of tibia of *T. vosmæri.*

PLATE XXVIII.

Fig. A. Side view of pelvis of *T. vosmæri*.

Fig. B. Side view of os ilium of *T. inepta*.

Fig. C. Side view of os ilium of *T. triserrata* (Newton Collection).

Fig. D. Radius of *Testudo triserrata* (No. 39950).

Fig. E. Radius of *T. vosmæri* (76. 11. 1. 119).

Fig. F. Ulna of *T. vosmæri* ♂ (76. 11. 1. 20).

Fig. G. Ulna of *T. vosmæri* ♀ (76. 11. 1. 120).

Fig. H. Posterior and end views of fibula of *T. triserrata* (No. 39962).

All these figures are of the natural size.

PLATE XXIX.

Femur (nat. size) of:—

Fig. A. *T. triserrata* (No. 39957).

Fig. B. *T. leptocnemis* (76. 11. 4. 11).

Fig. C. *T. vosmæri* (76. 11. 1. 116).

PLATE XXX.

Fig. A. Three views of carapace of *Testudo elephantopus* [specimen in the Oxford Museum], $\frac{1}{6}$ nat. size.

Fig. B. Three views of carapace of *Testudo nigrita*, $\frac{1}{6}$ nat. size.

PLATE XXXI. & XXXI. B.

Fig. A. Three views of *Testudo vicina*.

Fig. B. Lower view of the carapace of *Testudo ephippium*.

Fig. C. Three views of *Testudo nigrita* juv., from the typical specimen in the Collection of the Royal College of Surgeons.

All these figures are $\frac{1}{6}$ of the natural size.

PLATE XXXII.

Testudo microphyes, adult ♂.   Side view of head, nat. size.

PLATE XXXIII.

Testudo microphyes, adult ♂.   Upper view of head, nat. size.

PLATE XXXIV.

Testudo microphyes, ad. ♂, $\frac{1}{6}$ nat. size.   Lateral and upper views.

PLATE XXXV.

Fig. A. *Testudo microphyes*, adult ♂.   Lower view, $\frac{1}{6}$ nat. size.

Fig. B. *Testudo microphyes*, adult ♀.   Lower view, $\frac{1}{4}$ nat. size.

PLATE XXXVI.

Testudo microphyes, adult ♀.   Lateral and upper views, ¼ nat. size.

PLATE XXXVII.

Testudo microphyes, adult ♀ (type).   Three views of carapace, ¼ nat. size.

PLATE XXXVIII.

Testudo microphyes, half-grown ♂ (spec. c).   Three views of carapace.

PLATE XXXIX.

Testudo ephippium [from the typical specimen in the Museum of Science and Arts, Edinburgh], ⅒ nat. size.

PLATE XL.

Testudo abingdonii, adult ♂.   Lateral and front views, ⅙ nat. size.

PLATE XLI.

Testudo abingdonii, adult ♂.   Upper and lower views, ⅙ nat. size.

PLATE XLII.

Upper views, of the natural size, of the skulls of:—

Fig. A. *Testudo elephantopus* ♂.
Fig. B. *Testudo microphyes* ♀.
Fig. C. *Testudo ephippium* ♂.
Fig. D. *Testudo nigrita* ♂.
  a. Posterior margin of paroccipital.
  c. Tuberosity for the insertion of a portion of the temporal muscle.
  d. Groove separating the tuberosity from the zygomatic arch.

PLATE XLIII.

Lateral views, of the natural size, of the skulls of:—

Fig. A. *Testudo elephantopus* ♂.
Fig. B. *Testudo microphyes* ♀.
Fig. C. *Testudo ephippium* ♂.
Fig. D. *Testudo nigrita* ♂.
  e. Notch for the passage of the Eustachian tube.

PLATE XLIV.

Lower views, of the natural size, of the skulls of:—

Fig. A. *Testudo elephantopus* ♂.
Fig. B. *Testudo microphyes* ♀.
Fig. C. *Testudo ephippium* ♂.
Fig. D. *Testudo nigrita* ♂.

　　　a, c, d, e as in Plates XLII. & XLIII.
　　　b. Hollow in front of the occipital condyle.
　　　f. Outer pterygoid-edge.

PLATE XLV.

Figs. A–C. Skull of *Testudo microphyes* ♂. Lower and side views and articular surface of mandible.
Figs. D–F. Skull of *Testudo abingdonii* ♂. Lower and side views and articular surface of mandible.

PLATE XLVI.

Fig. A. Upper and lateral views of the atlas of *Testudo elephantopus*.
　　　a. Centrum.
Fig. B. Upper and lateral views of the atlas of *Testudo vicina* (centrum lost).
Fig. C. Three views of the fifth cervical vertebra of *Testudo elephantopus*.
Fig. D. Three views of the sixth cervical vertebra of *Testudo elephantopus*.
　　　All the figures are of the natural size.

PLATE XLVII.

Fig. A. Lateral view of the skull of *Testudo vicina*.
Fig. B. Three views of the seventh cervical vertebra of *Testudo elephantopus*.
　　　a. Summit of neural crest.
　　　b. Hollow behind the anterior zygapophyses.
Fig. C. Three views of the seventh cervical vertebra of *Testudo vicina*.
　　　a. Bifurcate summit of the neural crest.
　　　All these figures are of the natural size.

PLATE XLVIII.

Three views of the second (A), third (B), and fourth (C) cervical vertebræ of *Testudo abingdonii*, nat. size.

PLATE XLIX.

Three views of the fifth (D) and sixth (E) cervical vertebræ of *Testudo abingdonii*, nat. size.

PLATE L.

Three views of the seventh (F), and five of the eighth (G) cervical vertebræ of *Testudo abingdonii*, nat. size.

PLATE LI.

Fig. A. Front view of the humerus of *Testudo elephantopus*.

Fig. A'. Back view of the same.

Fig. B. Front view of the humerus of *Testudo ephippium*.

Fig. B'. Back view of the same.

        a. Ulnar tuberosity.

        b. Radial canal for blood-vessels.

           These figures are two thirds the natural size.

PLATE LII.

Three views, of two thirds the natural size, of the pelvis of *Testudo elephantopus*.

Fig. A. Front view.

Fig. B. Side view.

Fig. C. Top view.

        a. Symphysial bridge between the obturator-foramina.

        b. Lower portion of the pubic bones.

        c. Styliform process of the pubic bones.

        d. Posterior part of the ossa ischii.

PLATE LIII.

Figs. A, A'. Front and side views of the femur of *Testudo elephantopus*.

Fig. A''. Top view of the same.

        a & *b*. The confluent larger and lesser trochanters.

Fig. B. Top view of the femur of *Testudo ephippium*.

        a. Larger trochanter, separated by a wide groove from

        b. Lesser trochanter.

Fig. C. Scapulary of *Testudo elephantopus*.

Fig. C'. Another view of the upper portion, to show the regular position of the coracoid and acromion.

Fig. D. Carpus of *Testudo elephantopus*.

        u. Ulna.

        r. Radius.

        a. Coalesced scaphoid and os intermedium.

        b. Coalesced two radial ossicles of distal carpal series.

           All these figures are two thirds the natural size.

PLATE LIV.

Fig. A. Top view of the pelvis of *Testudo ephippium*.

Fig. B. Forearm of *Testudo ephippium*

     u. Ulna.   *r.* Radius.

Fig. C. Scapulary of *Testudo vicina*.

Fig. C'. Another view of the upper portion, to show the relative positions of the coracoid and acromion.

Fig. D. Forearm of *Testudo vicina*.

     u. Ulna.   *r.* Radius.

(Plates XXXVII., XXXIX., XLII.–XLIV., and LI.–LIV. are reprinted, without
alteration, from the 'Philosophical Transactions' of the Royal Society, vol. clxv.
part i.)

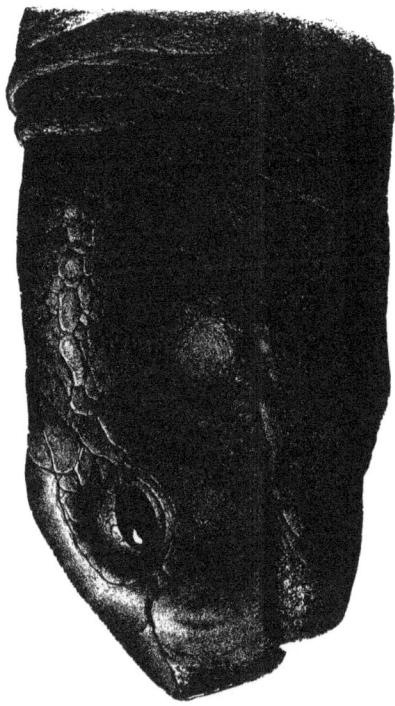

Pl. II.

Testudo elephantina.
(Aldabra.)

Pl. III

C. L. Griesbach

Minterr freres imp

Testudo elephantina ♂.

Pl. IV

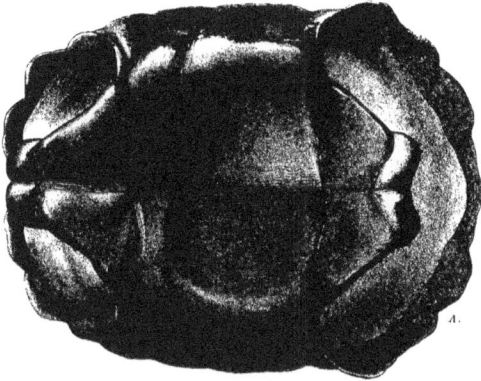

A.

B.

G. L. Griesbach

Xitaaff. Hrsg. img.

A. Testudo elephantina.   B. Testudo daudinii.

Testudo daudinii ?

Pl. VI.

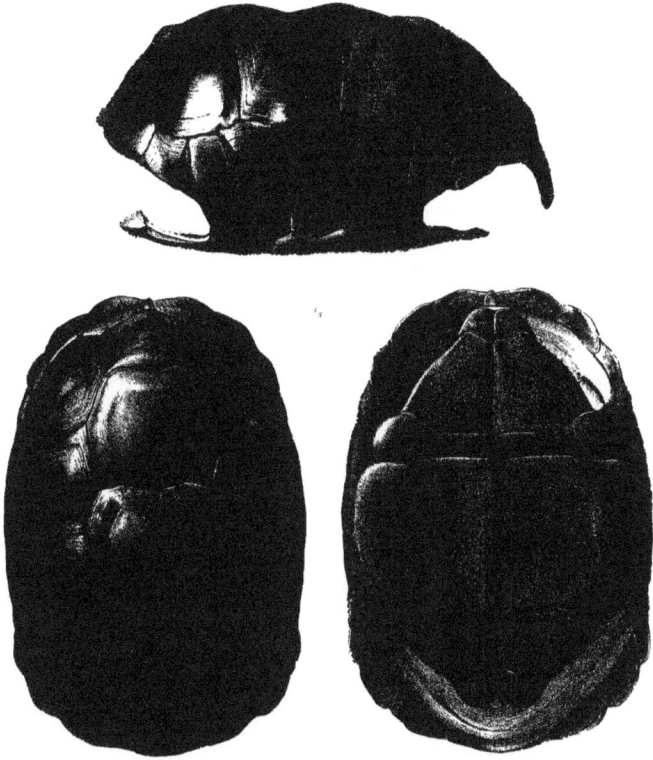

Testudo ponderosa. ♀.

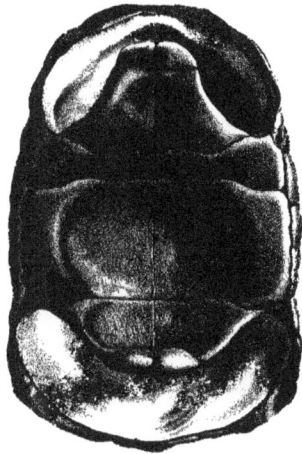

Testudo holotissa.

Pl. VIII.

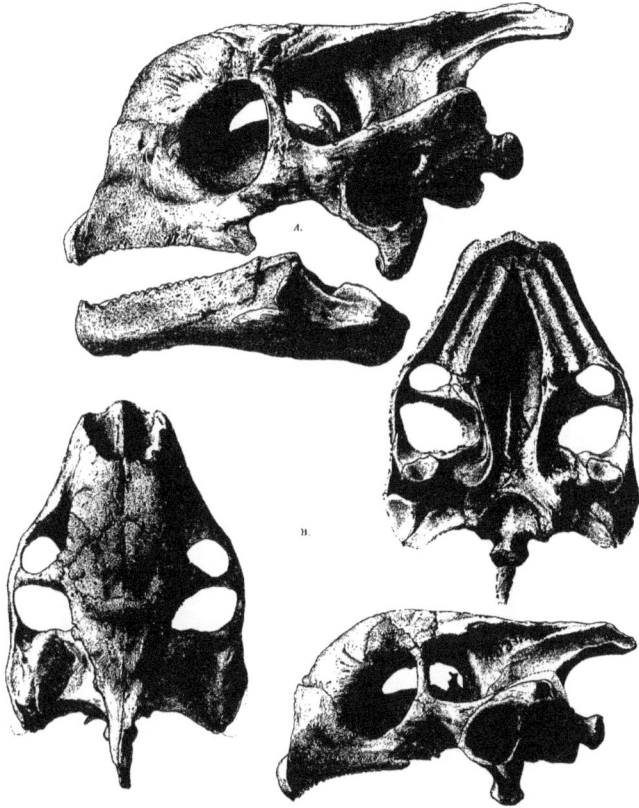

Greenich. Mintern Bros. imp.

A. Testudo elephantina   B. Testudo ponderosa.

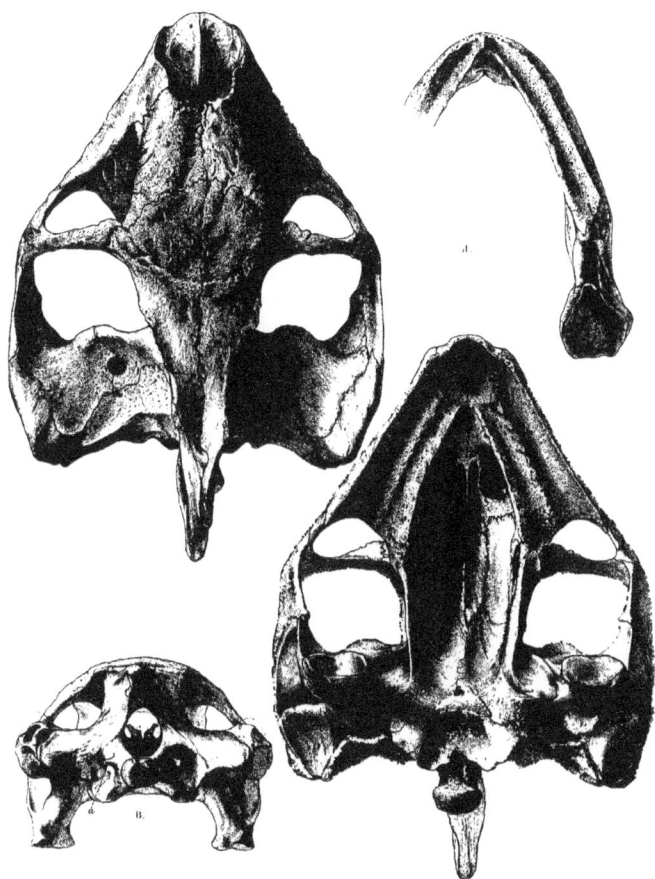

C. I. Griesbach.

A. *Testudo elephantina*.   B. *Testudo ponderosa*.

Pl. X.

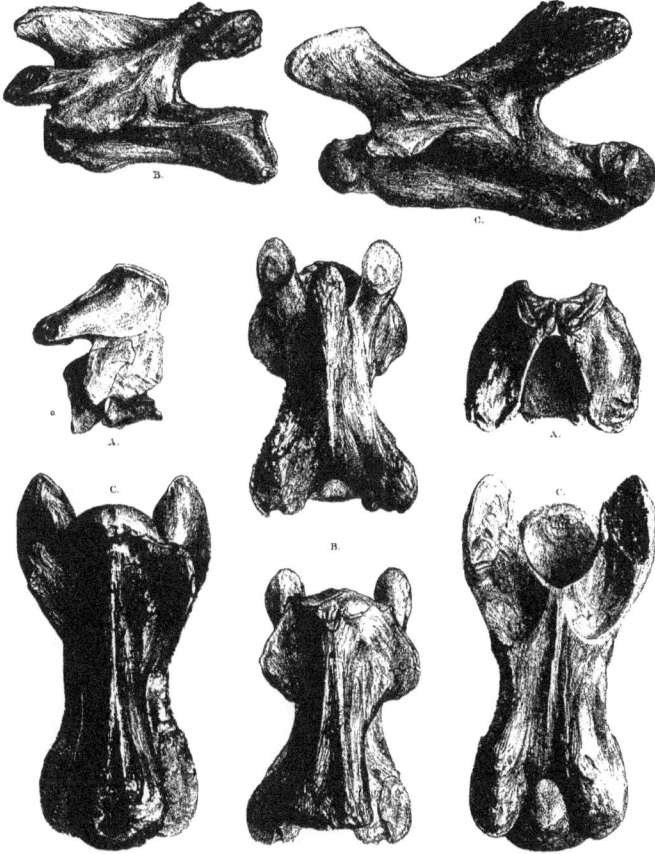

Cervical vertebræ of Testudo elephantina.
A. First. B. Second. C. Third.

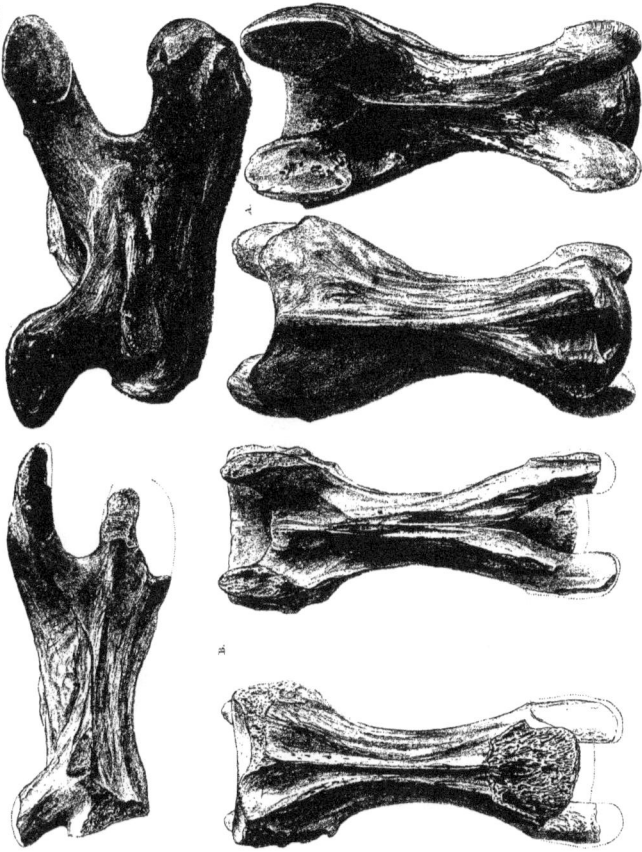

Pl. XI.

Fifth cervical vertebra.

A. Testudo elephantina.   B. Testudo vosmaeri

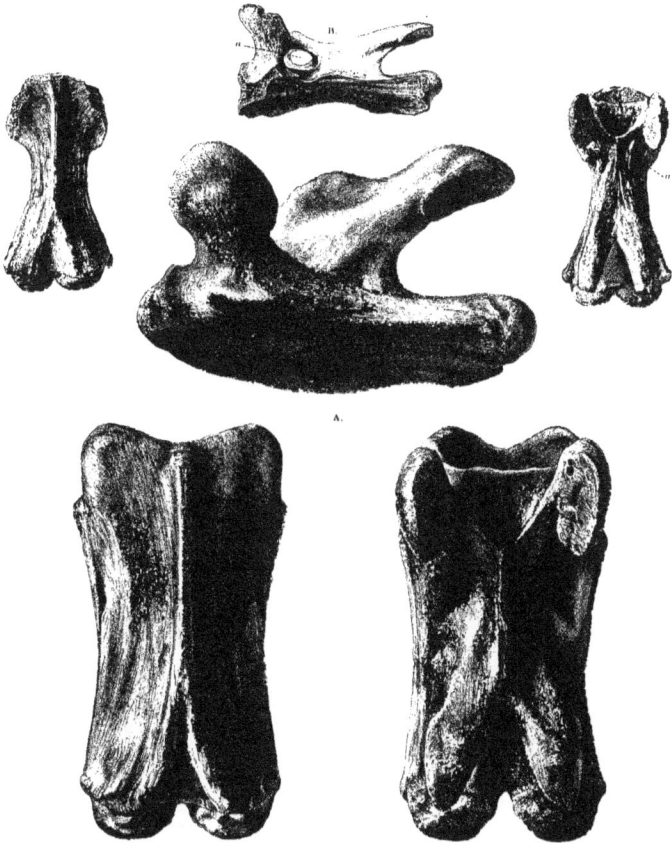

Sixth cervical vertebra,
A. *Testudo elephantina.*   B. *Testudo vosmari.*

Pl. XIII

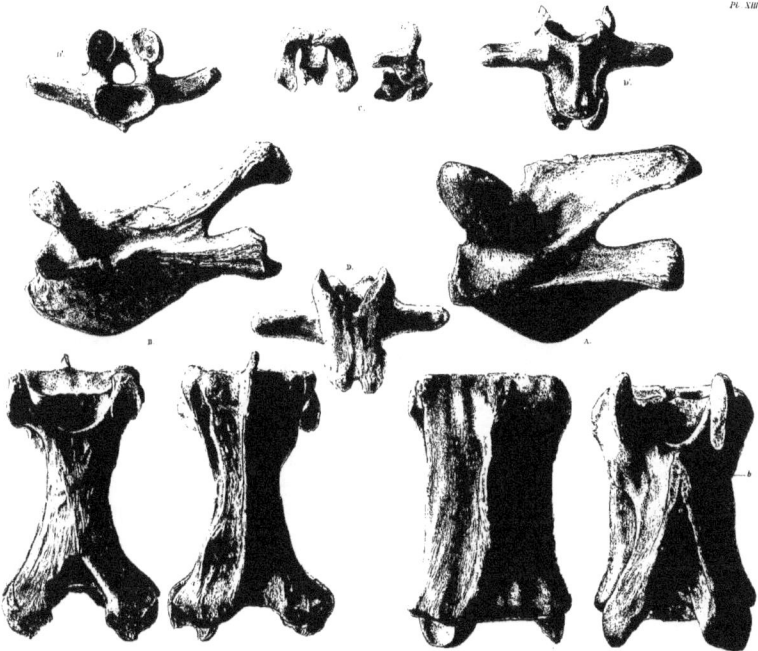

A. Sixth cervical vertebra of Testudo elephantina, and B. of Testudo vosmaeri.
C. Atlas of Testudo ponderosa, D. Sixth caudal vertebra of Testudo elephantina.

Mintern Bros. imp.

8. Miniaturen als ia hock

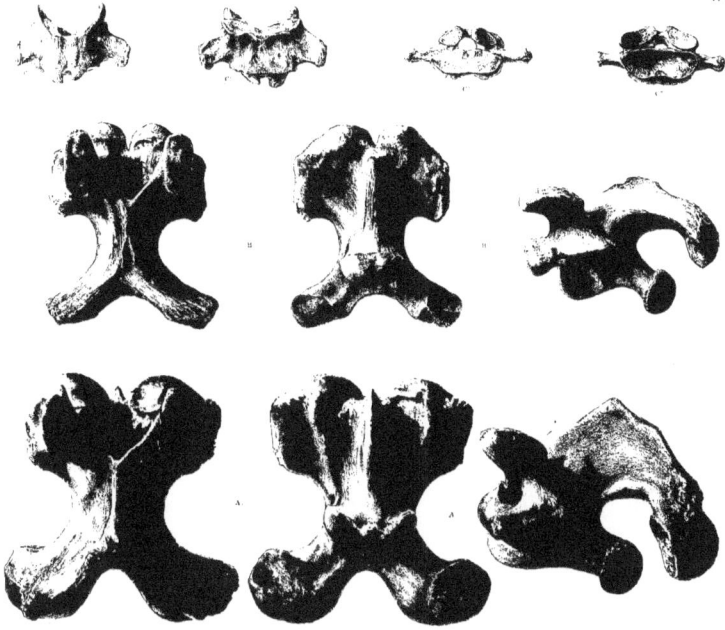

A & B. Right cervical vertebra of Testudo elephantina (A) and Testudo vosmaeri (B).
C. Eighteenth caudal vertebra of Testudo elephantina.

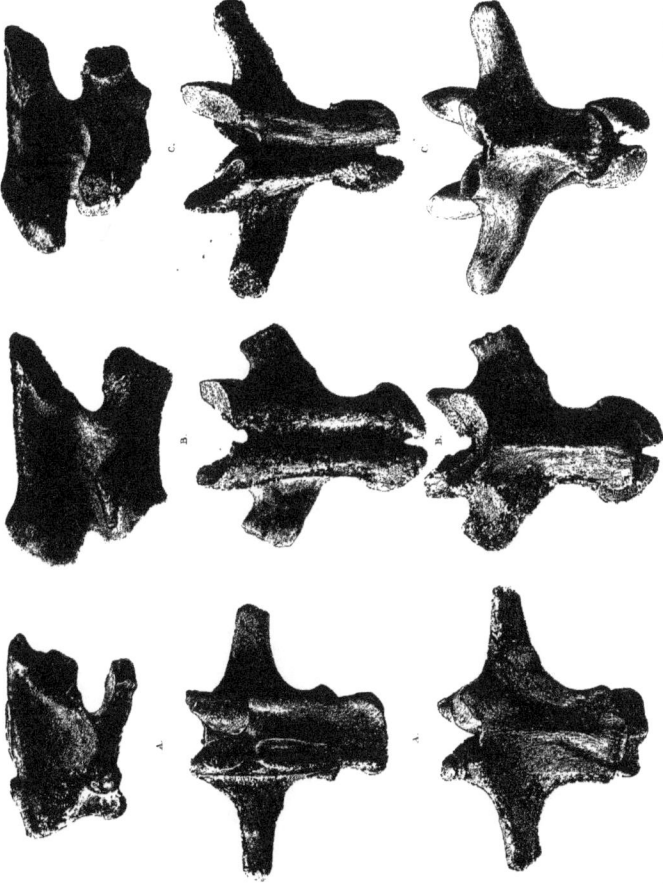

Pl. XV.

Caudal vertebræ of Testudo elephantina.
A. First.   B. Fourth.   C. Sixth.

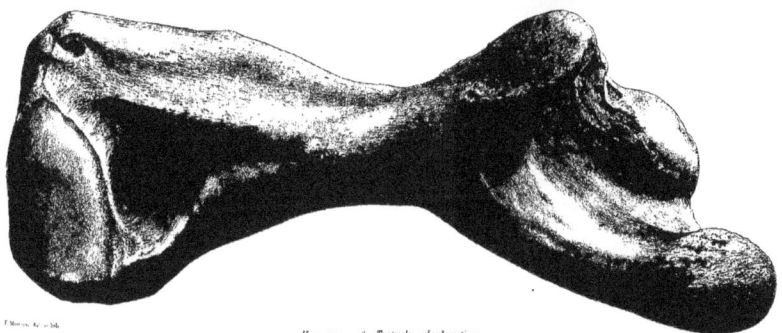

Pl. XVI

Humerus of Testudo elephantinus.

T.Marsten del. et lith.   Martens, Pere imp.

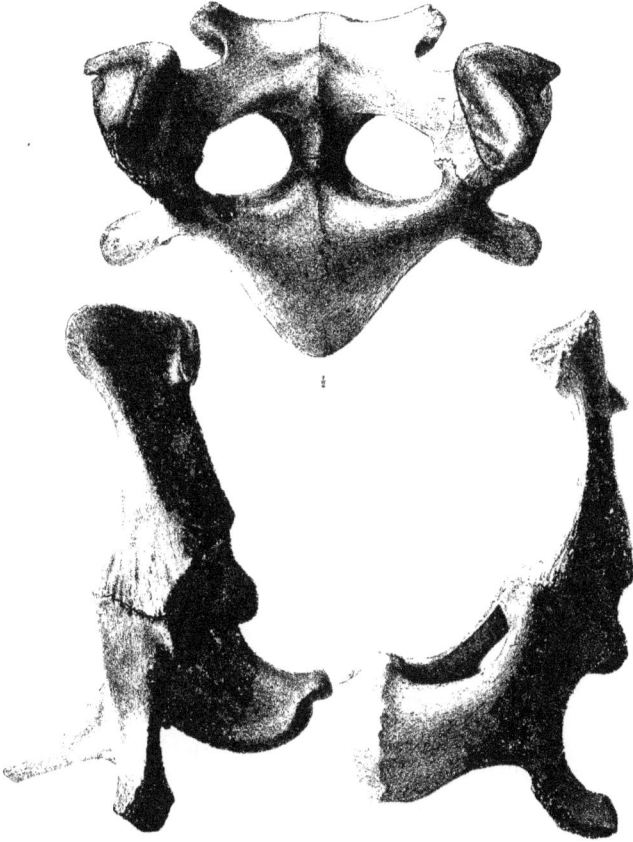

R.Mintern del et lith.   Mintern Bros. imp.

Pelvis of Testudo elephantina.

Pl. XVIII.

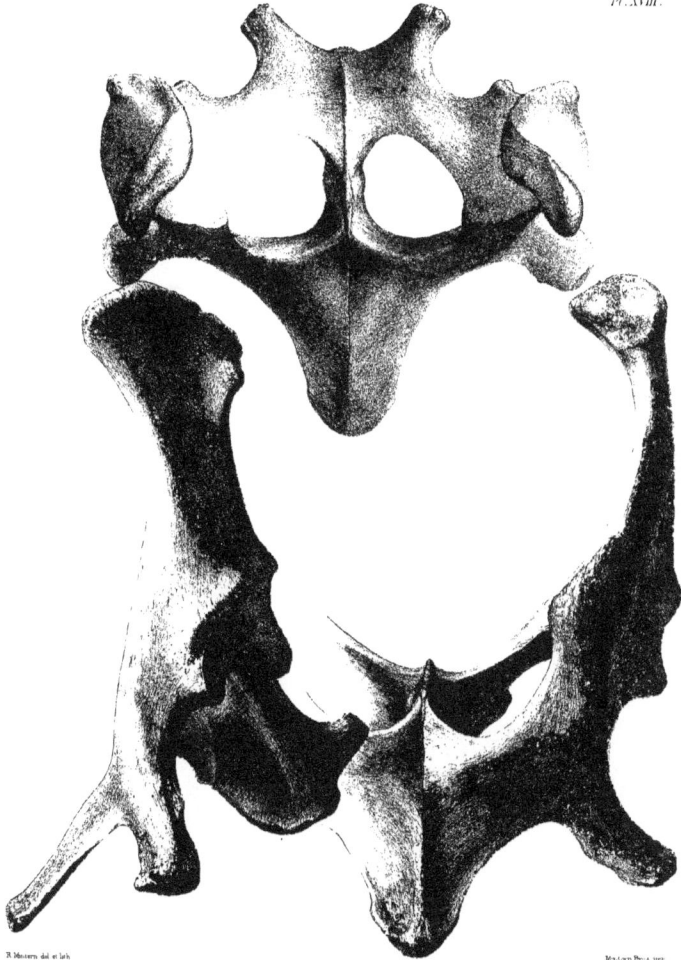

Pelvis of Testudo ponderosa.

Pl. XIX.

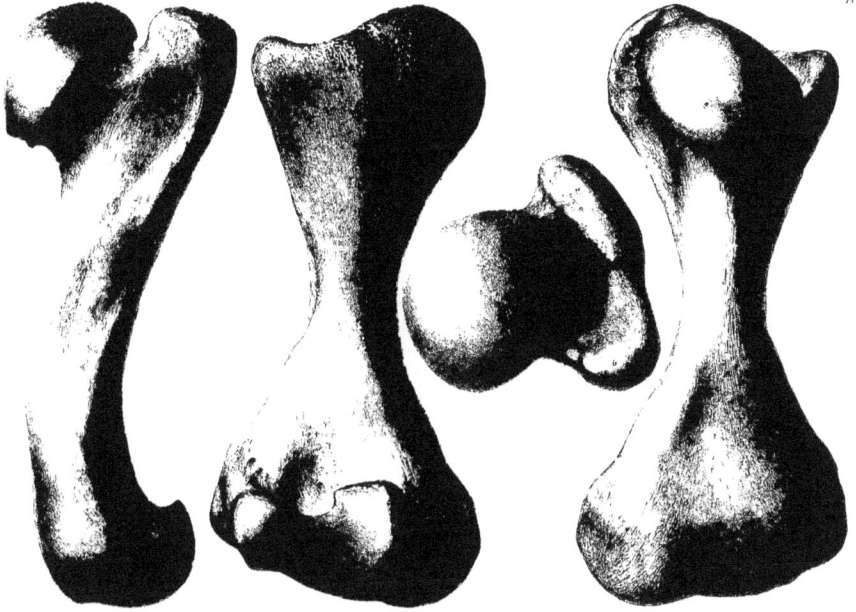

Femur et Testudo elephantina.

Imprim. Bies imp.

Pl. XV.

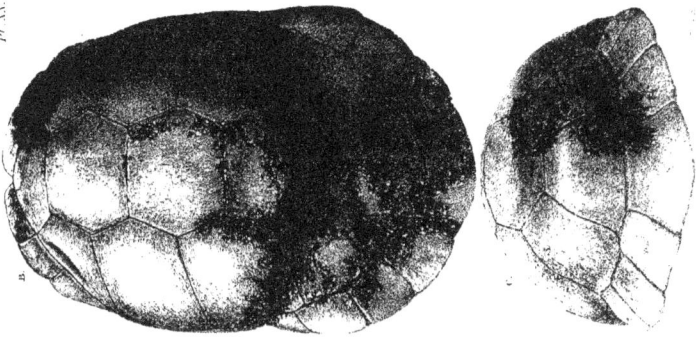

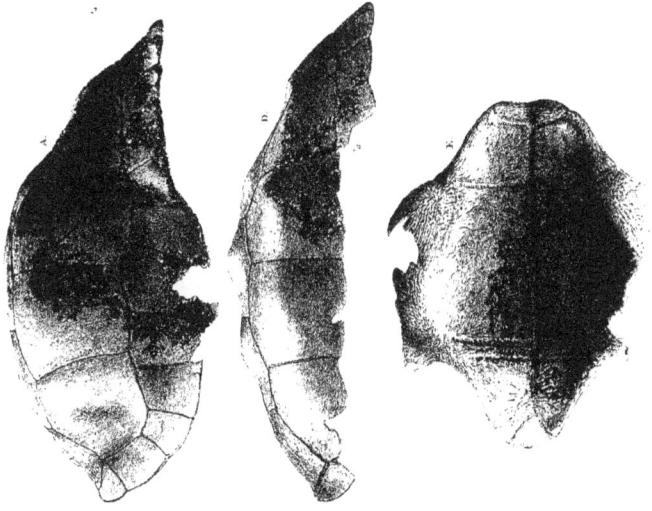

Pl. XXI.

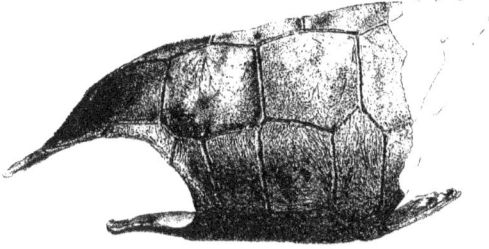

3⁄5

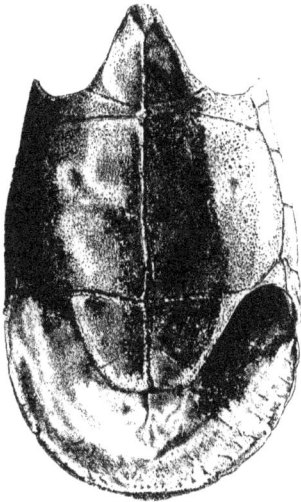

Testudo vosmæri, ♂.
(Rodriguez.)

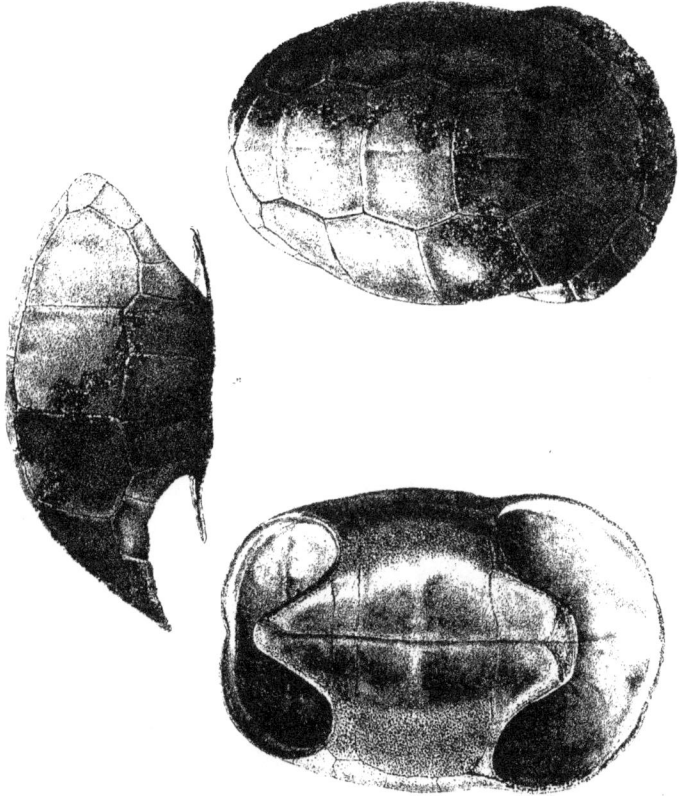

R. Norman del. et

Pl. XXIII

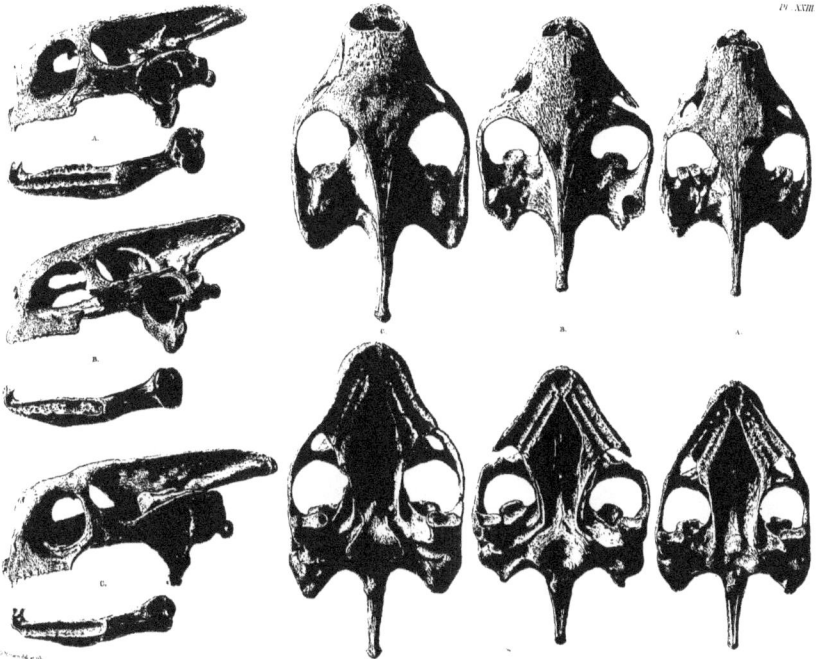

Skull.
A. Testudo trinervata. B. Testudo inepta. C. Testudo vosmaeri.

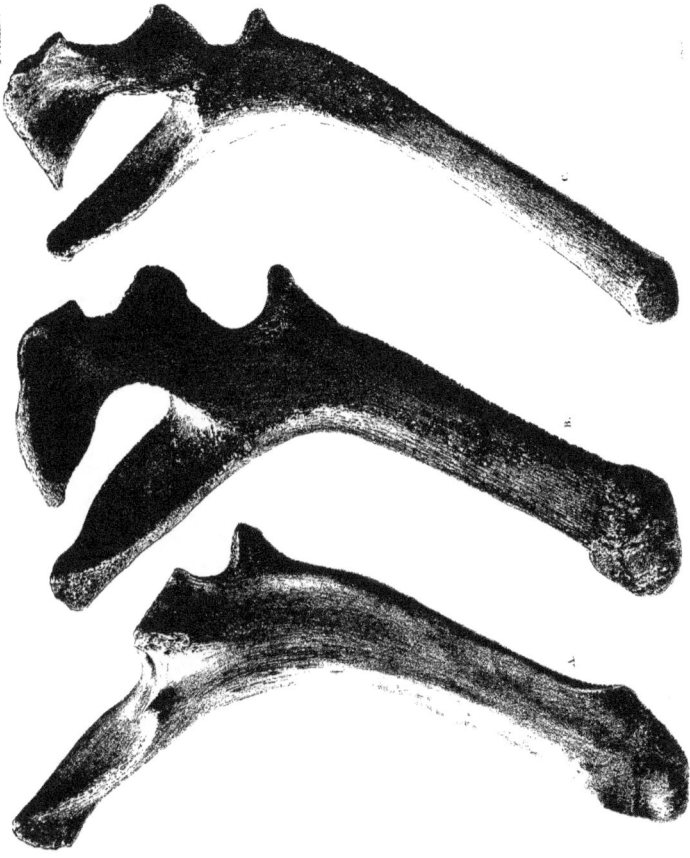

Pl. XXV.

A. *T. podotores.*   B. *T. inserrata.*   C. *T. vramers.*

Scapula.

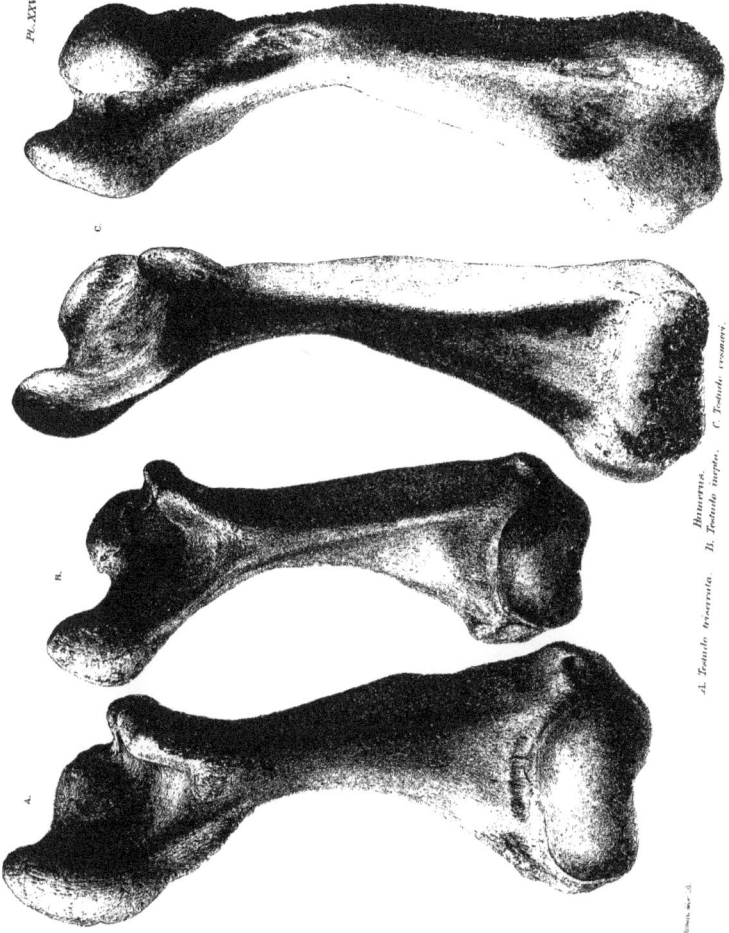

Pl. XXV

Humerus.
A. Testudo triserrata.   B. Testudo incepta.   C. Testudo ecaudavi.

Pelvis.
A. Testudo triserrata.   B. Testudo inepta.   C. Testudo vosmaeri

Pl. XXVII.

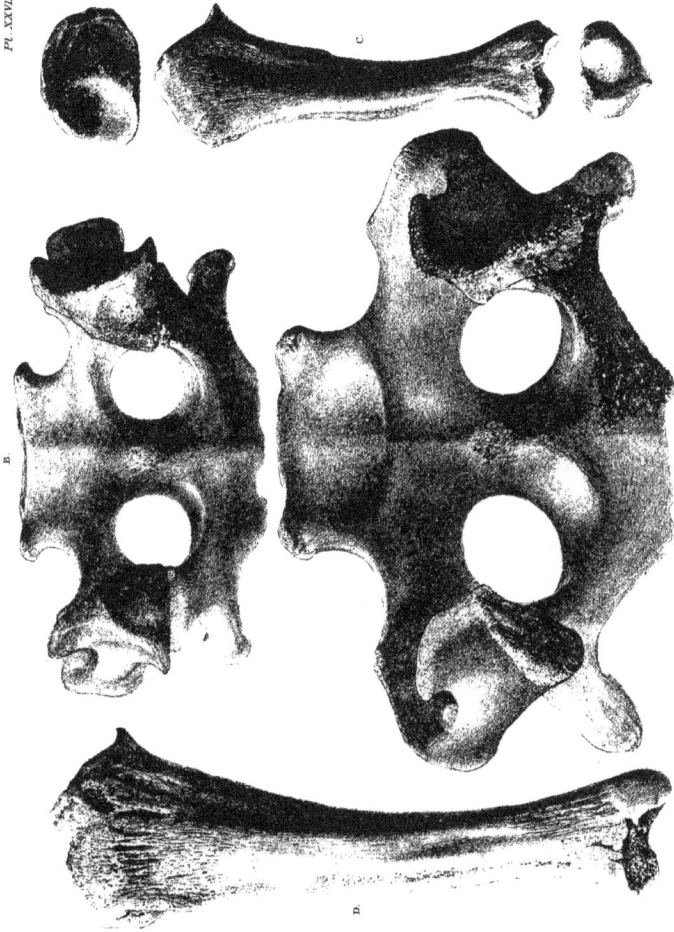

A. Pelvis of Testudo inepta, and B. of Testudo vosmeri.
C. Tibia of Testudo inserrata, and D. of Testudo vosmeri.

Kochari, Beret imp

B. Marion del. et lith.

R.Mintern del.et lith.

Mintern Bros imp

A. Pelvis of T.vosmari. B. Os ilium of T. inepta, and C. of T. triserrata. D. Radius of T. triserrata and
E. of T.vosmari. F. & G. Ulna of T. vosmari ♂ & ♀. H. Fibula of T. triserrata.

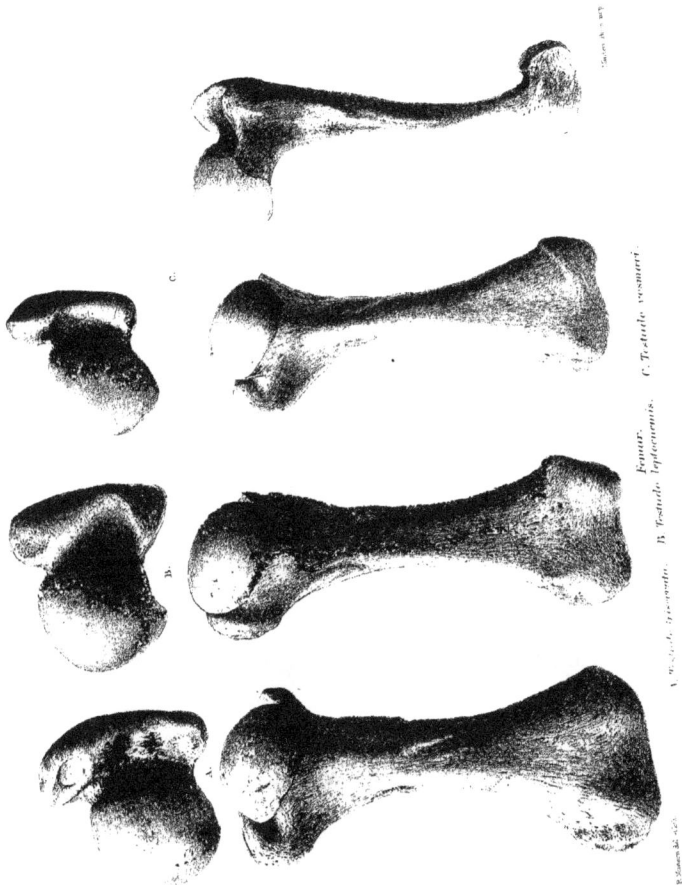

A. Testudo argentina. B. Testudo leptocnemis. C. Testudo vesanus.

Femur.

B

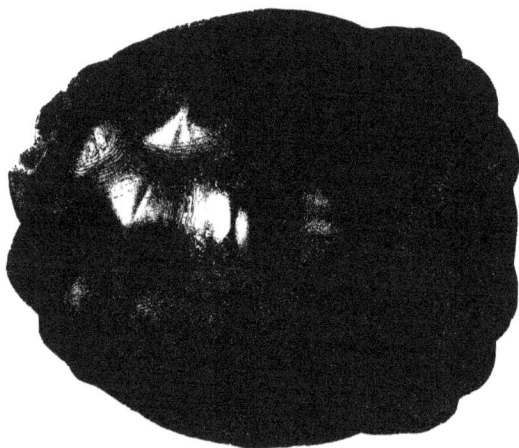

A. Testudo elephantopus
(Galapagos)

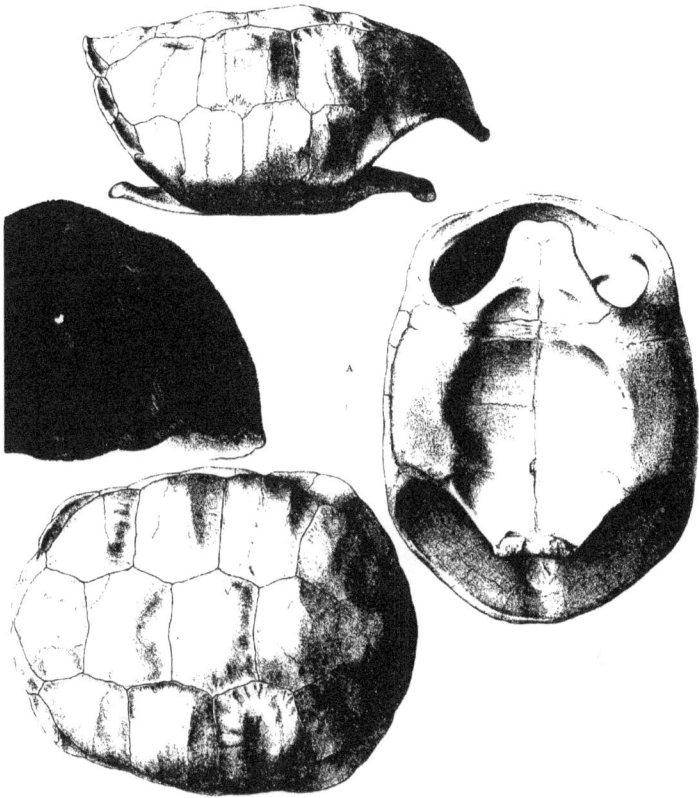

A

B. *Testudo nigrita*.

Mintern Bros. imp.

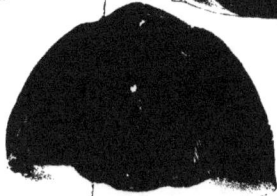

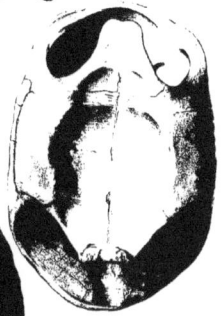

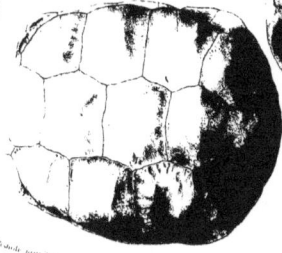

A. *Testudo elephantopus*
(Galap: Isid:)   B. *Testudo nigrita*

Pl. XXXI.

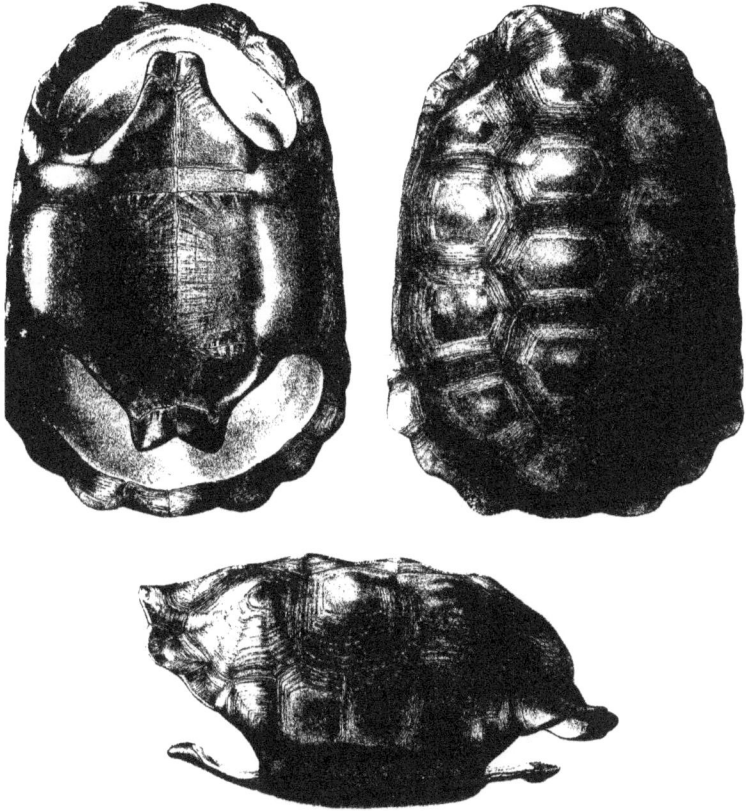

G. H. Ford.

Testudo vicina.
(South Albemarle.)

Hanhart, Imp.

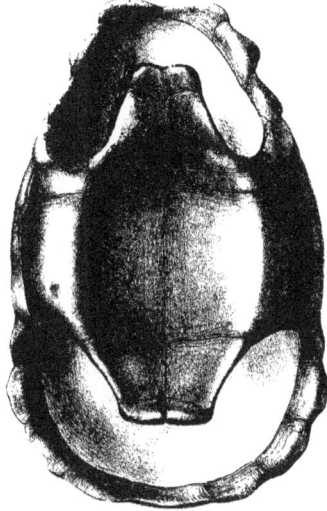

B.

C.

G H St t

Masson Paris imp

B. Testudo ephippium. C. Testudo nigrita.
(Galapagos isl.)

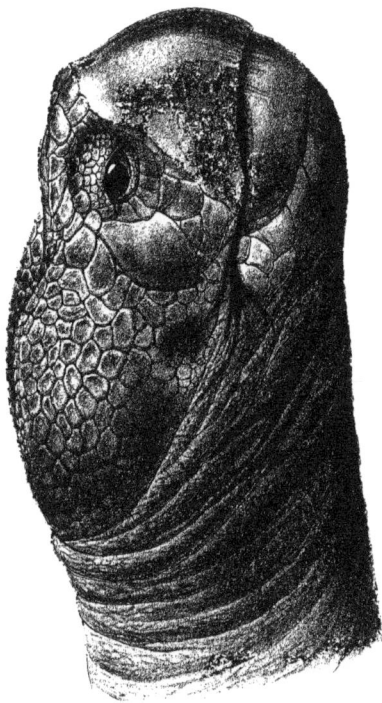

Pl. XXXIII.

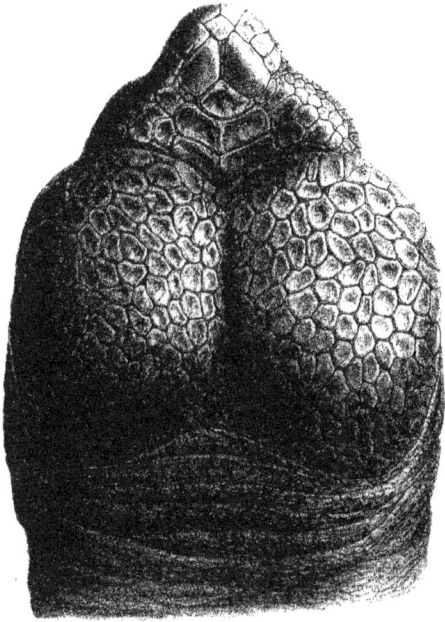

Testudo microphyes ad. ♂
,North Albemarle Isl.,

R. Mintern del. et lith.

Mus. red. soc. 1sq.

a.

Testudo microphyes, ad. ♂.
(North Albemarle)

B.Mintern del. et lith.   Mintern Bros. imp.

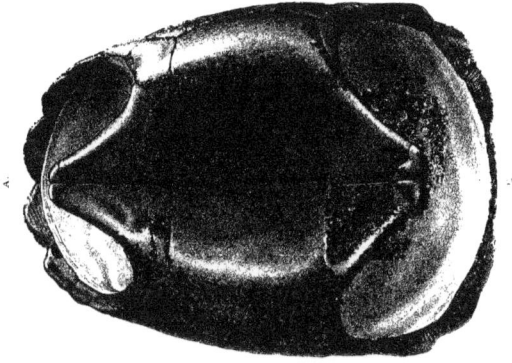

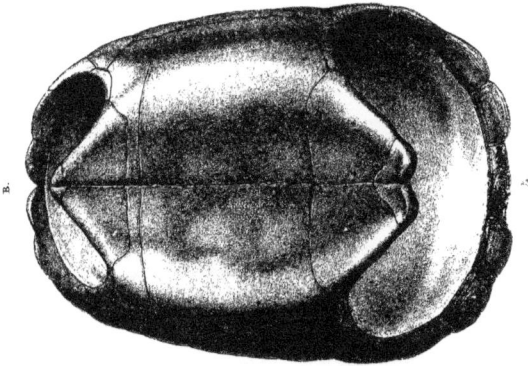

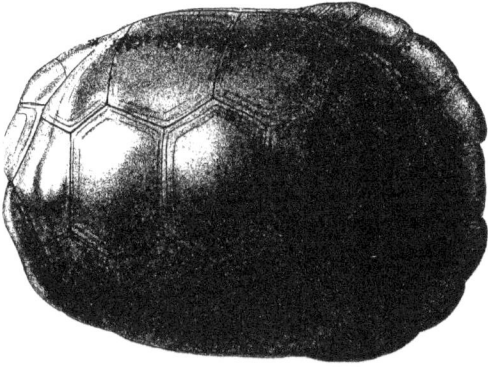

Pl. XXXVI.

Testudo microphyes. ad. ♀.
(North Albemarle.)

R. Mintern del et lith.   Mintern Bros. imp.

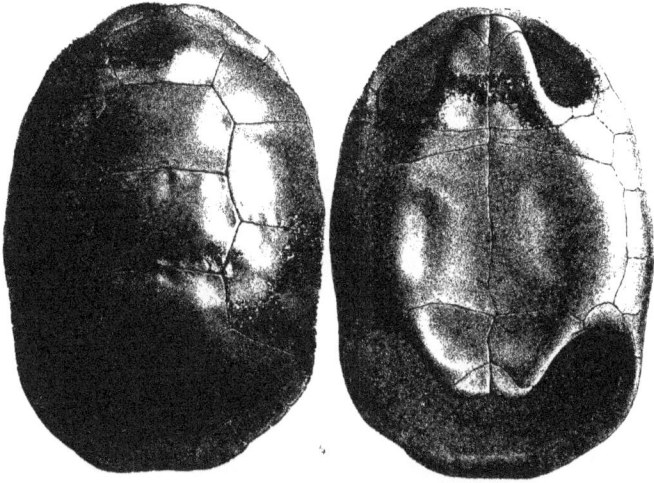

Pl. XXXVII.

W.H. Wesley

Mintern Bros. imp.

Testudo microphyes, ad. ♂ (type)
(North Albemarle Isl.)

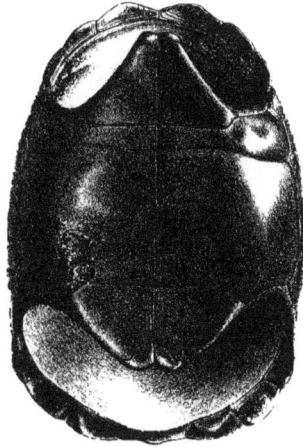

Pl. XXXVIII.

Testudo microphyes, hyr d.

Testudo ephippium, ad. ♂.
(Indefatigable Isl.)

G.H.Ford

Mintern Bros. imp.

R. Martens del. et lith.

Testudo abingdonii, ad. ?.
(Abingdon Isl.)

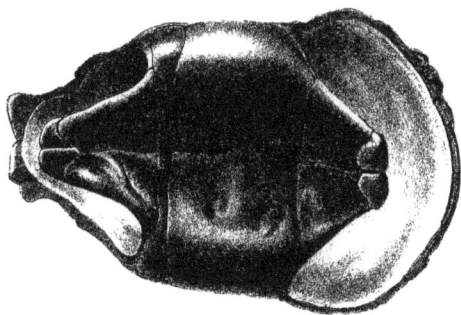

Pl. XLII.

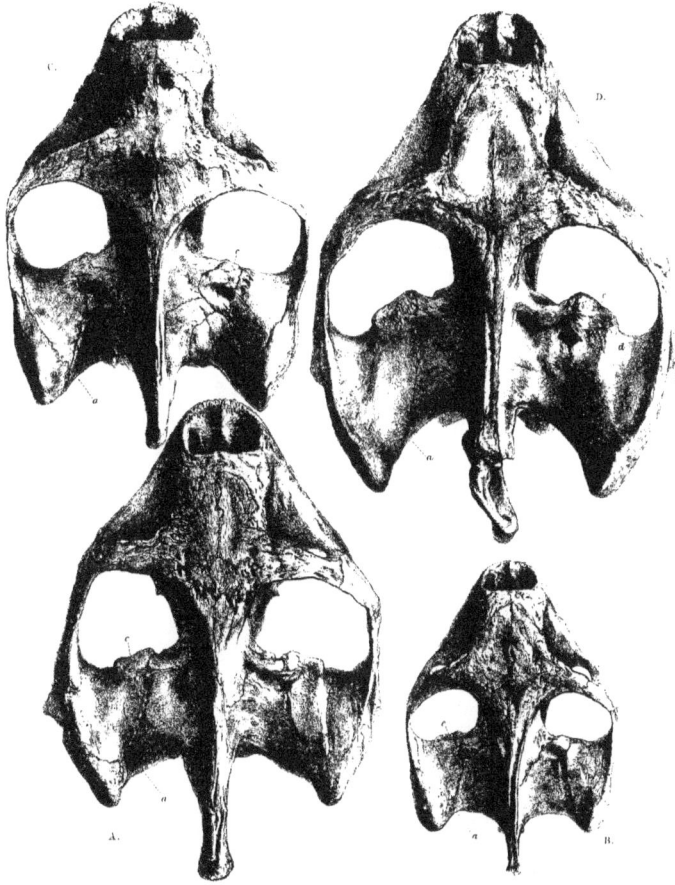

W.H.Wesley

A. *Testudo elephantopus* ♂.   B. *Testudo microphyes* ♀.   C. *Testudo ephippium* ♂.   D. *Testudo nigrita* ♂.

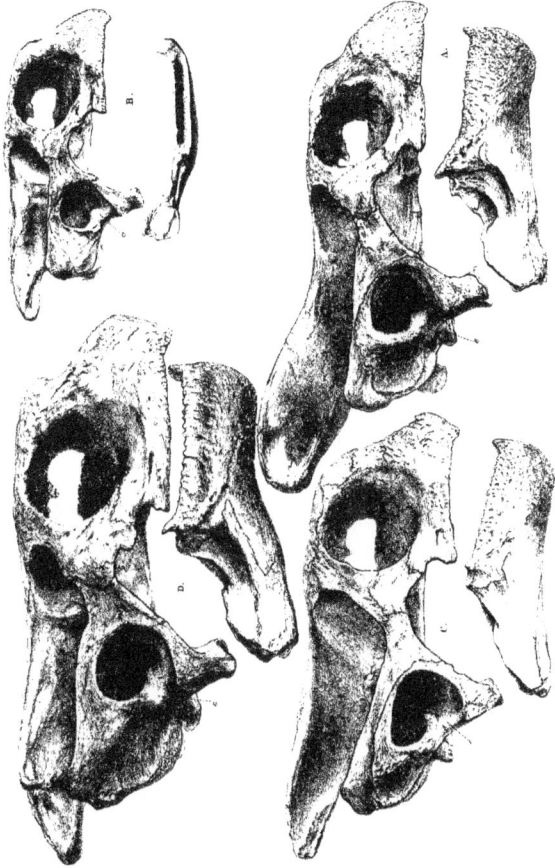

A. Testudo elephantopus ♂.   B. Testudo microphyes ♂. (Testudo ephippium ♀.   D. Testudo nigrita ♀.

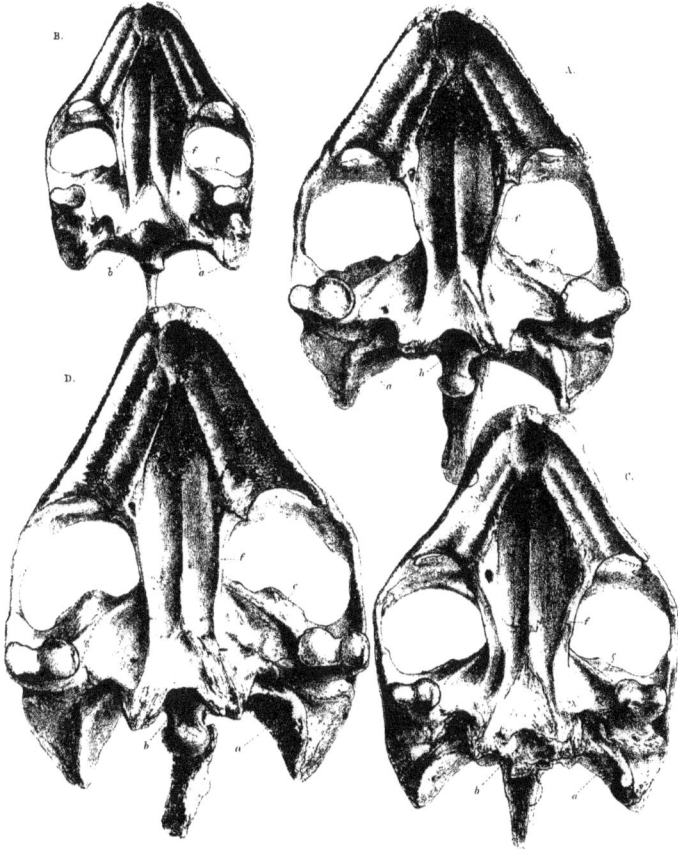

W.H.Wesley   Mintern Bros. imp.

A. Testudo elephantopus ♂.   B. Testudo microphyes ♀.   C. Testudo ephippium ♂.   D. Testudo nigrita ♂.

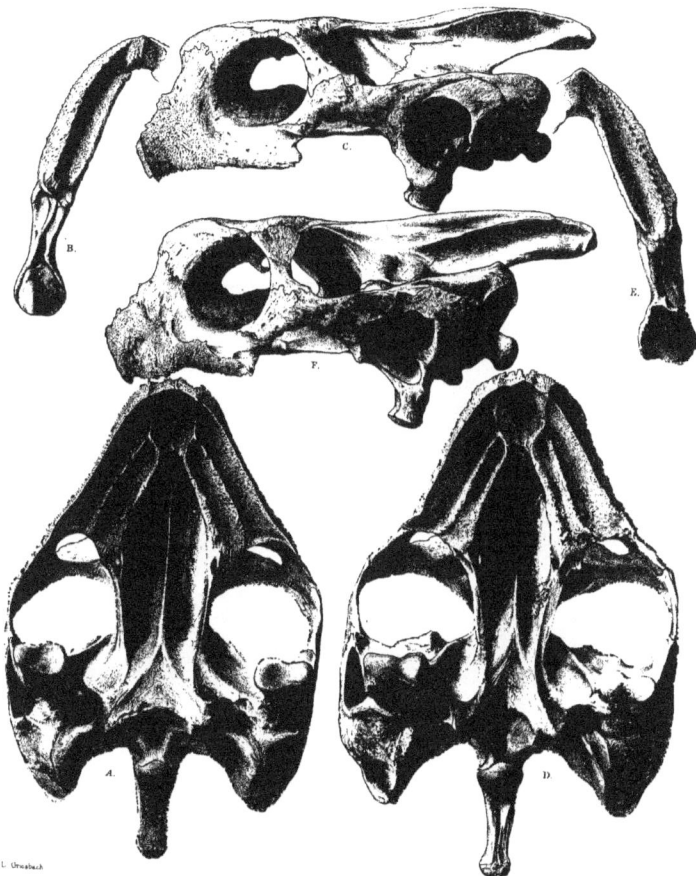

B.

C.

E.

F.

A.

D.

O. L. Ursabach.

Pl. XLVI.

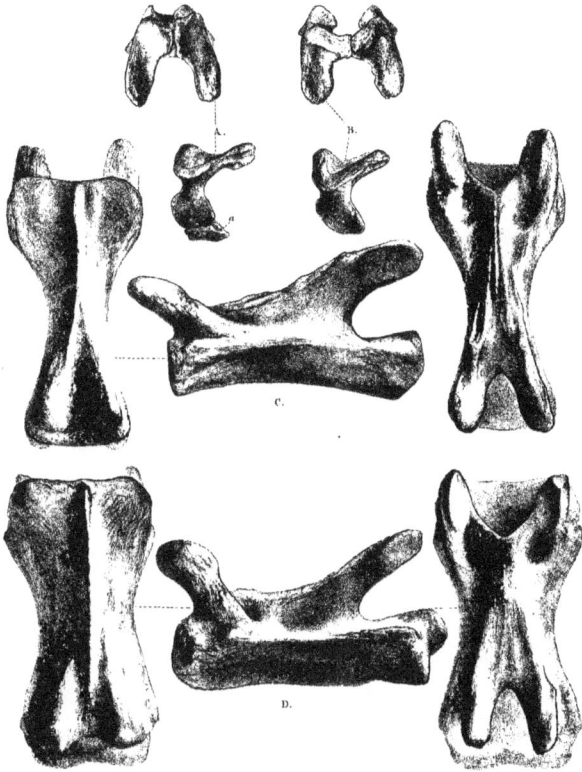

A. Atlas of Testudo elephantopus, and B. of Testudo vicina.
C. Fifth, and D. Sixth Cervical vertebra of Testudo elephantopus.

Pl. XLVII.

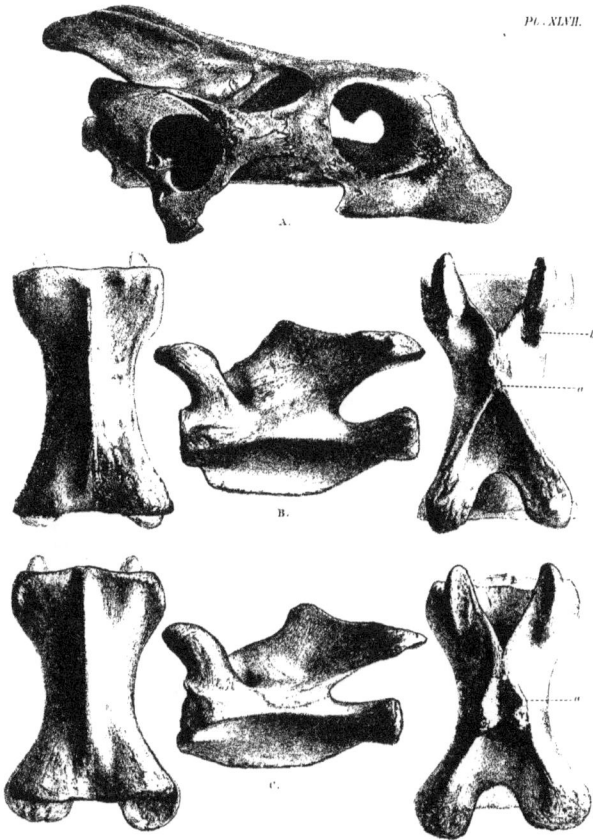

A. Skull of Testudo vicina.
B. Seventh cervical vertebra of Testudo elephantopus; and C. of Testudo vicina.

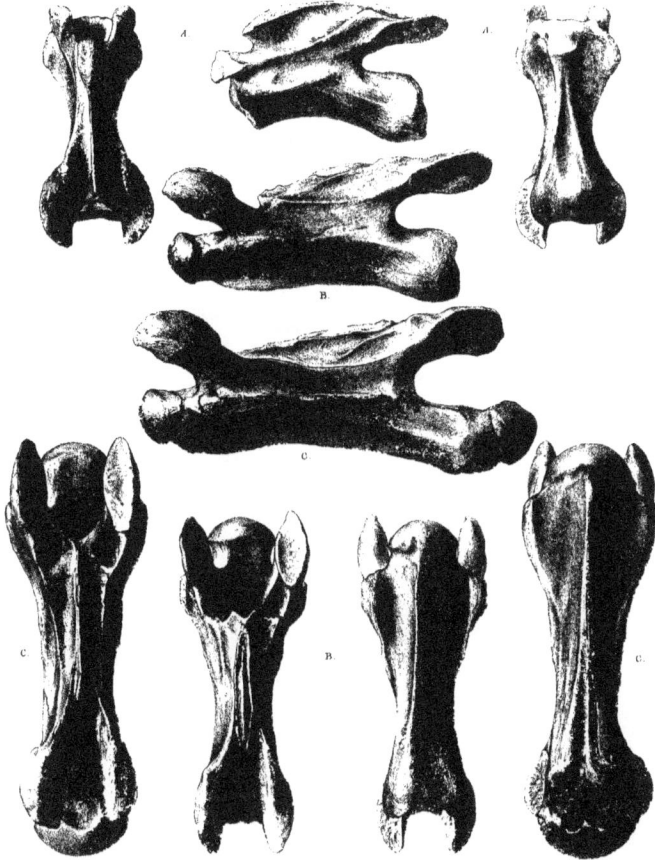

Pl. XLVIII.

Testudo abingdonii.
A. Second, B. Third, C. Fourth. cervical vertebra.

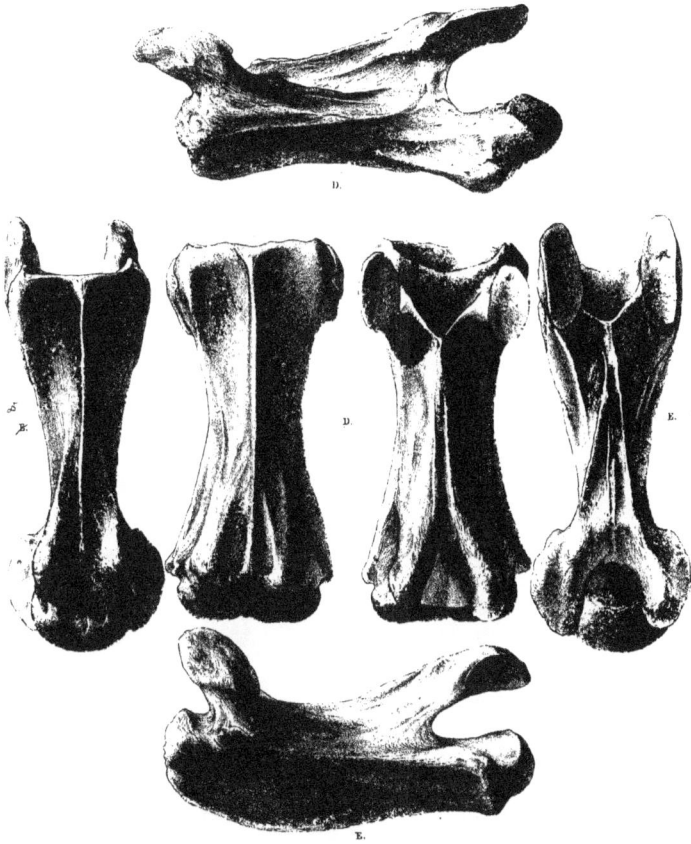

Pl. XLIX.

C.I. Griesbach

Testudo abingdonii.
D. Fifth, and E. Sixth cervical vertebra.

Mintern Bros. imp

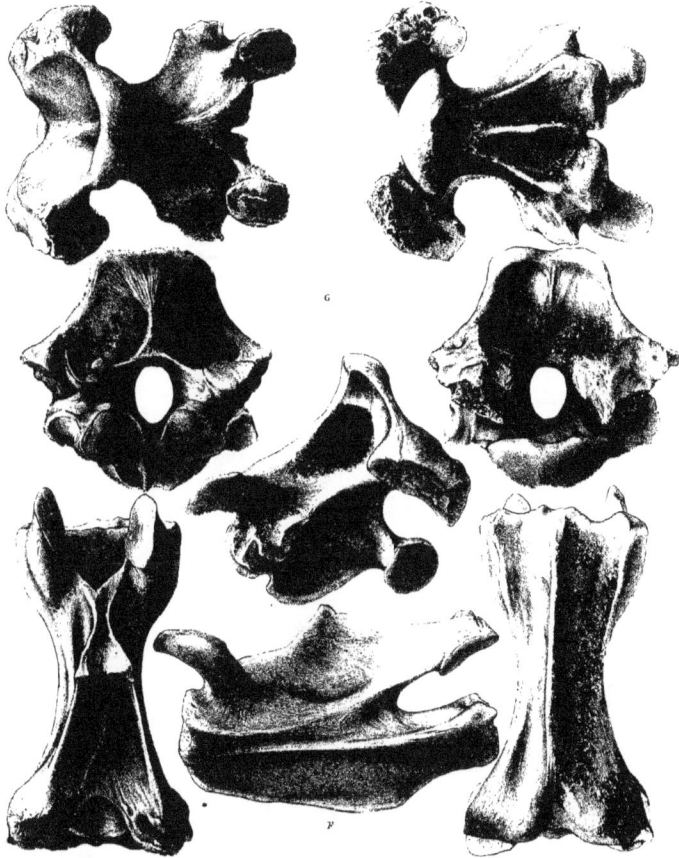

Pl. I.

C. I. Grönbach.

Testudo abingdonii.
F Seventh. and G. Eighth cervical vertebra.

Mintern Bros imp.

Pl. LI.

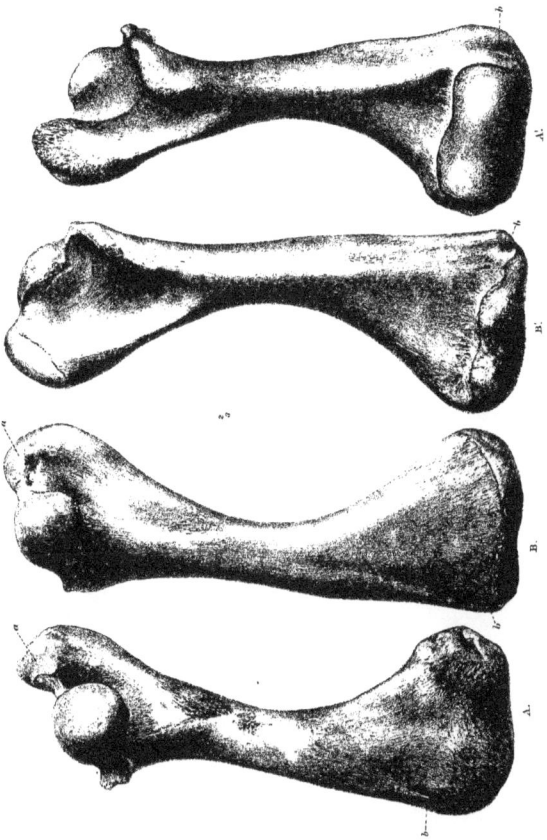

A'.

B'.

a

B

b

b

a

b

A.

Pl. LII.

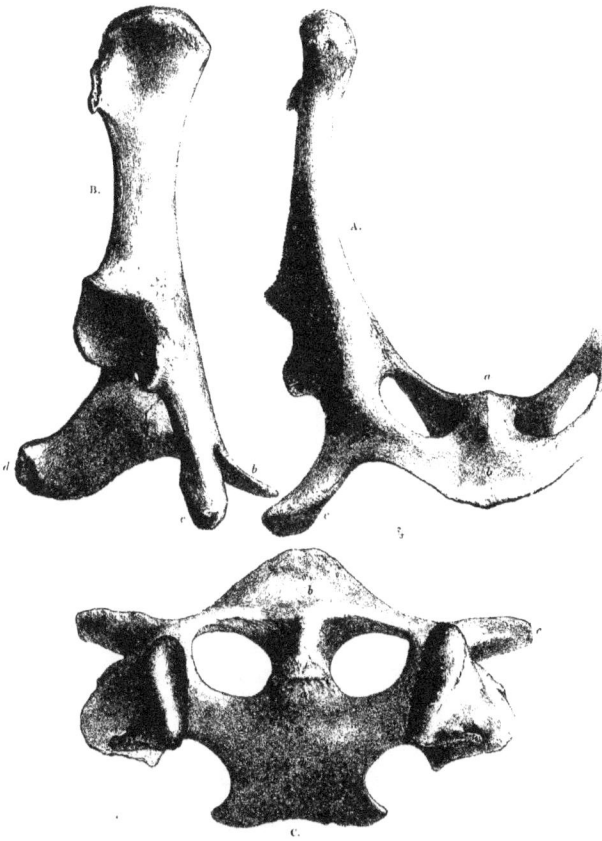

Pelvis d' Testudo elephantopus.

.

Pl. LIII.

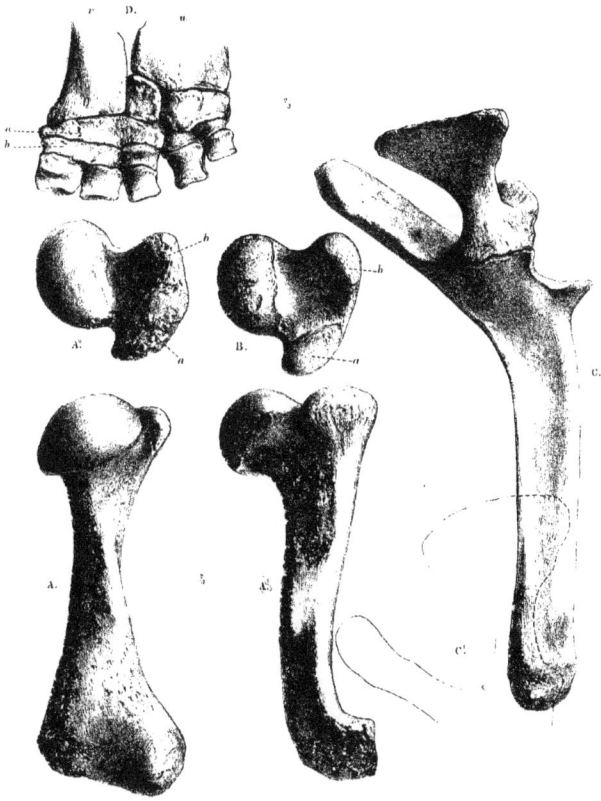

A. Femur of T. elephantopus and B. of T. ephippium. C. Scapulary and D. Carpus of T. elephantopus.

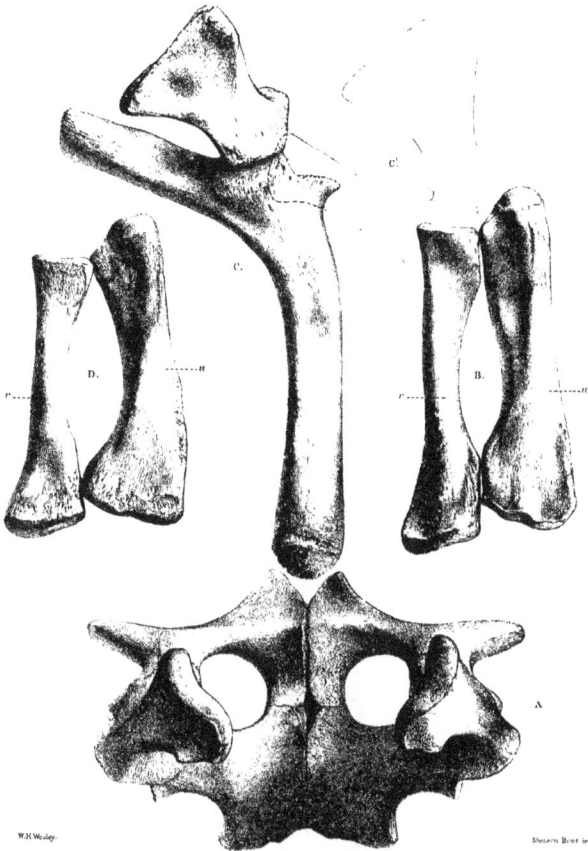

A. Pelvis of T. ephippium. B. Forearm of T. ephippium. C. Scapulary of T. vicina. D. Forearm of T. vicina.

www.ingramcontent.com/pod-product-compliance
Lightning Source LLC
Chambersburg PA
CBHW021941220326
41599CB00013BA/1483